Satellite Imagery Interpretation For Forecasters

Compiled and Edited by
Peter S. Parke, NWS

Volume 1

General Interpretation
Synoptic Analysis

The National Weather Association
501 Capitol Court, N.E., Suite 100
Washington, D.C. 20002-4937

First Printing December 1986
Second Printing May 1993

International Standard Book Number

Volume 1: 1-883563-01-1

Volume 2: 1-883563-02-X

Volume 3: 1-883563-03-8

3 Volume Set: 1-883563-04-6

All rights reserved. No part of this book may be reproduced or utilized in any form or by any means, electronic or mechanical, including photocopying, recording, or by any information storage and retrieval system, without permission in writing from the publishers.

Printed in the United States of America

PREFACE

During the past two decades satellite information has played an increasingly important role in weather analysis and forecasting. Governmental and private weather services, broadcast meteorologists, consultants, university based educators, and others have had to master all aspects of satellite interpretation from basic cloud identification to detailed assessments of atmospheric processes in both data-rich and data-poor regions. Until now, the only way satellite imagery users could acquire this information was to read various journals and technical reports they became aware of, attend scientific conferences, or participate in a limited number of seminars and workshops focusing on satellite imagery interpretation.

In 1985, the National Weather Service (NWS) and the National Environmental Satellite, Data and Information Service (NESDIS) began a cooperative effort to compile the "state-of-the-art" documentation on imagery interpretation for training and applied research purposes. That effort culminated in National Weather Service Forecasting Handbook Number 6 — "Satellite Imagery Interpretation For Forecasters." The Handbook was distributed throughout the National Weather Service and the weather and oceanography branches of the Department of Defense.

Following this, many non-governmental and governmental people expressed a desire to obtain the Handbook. The National Weather Association, as part of its overall philosophy, welcomed the opportunity to publish the Handbook and make it available, at low cost, to operationally-focused users worldwide.

This monograph is published in three volumes. The first includes information about satellite observing systems, basic imagery interpretation and synoptic analysis. The second focuses on precipitation and convection (severe and non-severe). Volume 3 addresses tropical weather; fog and stratus; atmospheric aerosols; and, winds and turbulence. Volume 3 also includes a comprehensive glossary.

NWA plans to coordinate with NWS and NESDIS to update this monograph on a periodic basis as new advances are made in the understanding of atmospheric processes and satellite imagery interpretation. The three volume strategy will facilitate usage and updating, and keep costs for obtaining future volumes at a more reasonable level.

We welcome your comments concerning the content of this Monograph set. NWA will provide feedback to NESDIS and NWS for enhancing the content of NWS Forecasting Handbook Number 6 and use the information to update this Monograph set, as well.

ACKNOWLEDGMENTS

A reference file of this nature is based on the work, expertise, and talents of many. The input by the people listed below is gratefully acknowledged.

National Weather Service

COORDINATOR: Michael T. Young,
    NWS Satellite
    Program Leader

Planning Research Corporation

PHOTOGRAPHIC RESEARCH: Laura Hogan,
    Meteorological Data
    File Manager

National Environmental Satellite, Data and Information Service

COORDINATOR: H. Michael Mogil,
    Chief, Satellite
    Training Branch

TECHNICAL ADVISOR: Gary Ellrod
    Meteorologist

PHOTOGRAPHIC SUPPORT: Gene Dunlap,
    Photographer

John Shadid,
    Visual Specialist

Air Weather Service

Col John H. Taylor
Col. John W. Oliver
Lt. Col. Dennis E. Bielicki
Lt. Col. Charles P. Guard
MSgt. Frederick E. Gesser

Special thanks to Ralph Anderson, Susan Holmes, and Mark Waters, all of NESDIS, for their assistance and support of this project. We also must thank all the authors whose publications are included in this Handbook. Without their efforts, this Handbook would never have been possible.

TABLE OF CONTENTS

Page

CHAPTER 1

GENERAL INTERPRETATION 1-1

    A. Introduction to Weather Satellite Imagery 1-A-1

    B. Reliability of Infrared GOES Data at High Latitudes 1-B-1

    C. Displacement Error of Satellite Cloud Tops 1-C-1

    D. Cloud Location Corrections Near the Horizon 1-D-1

    E. Supplement, including NOAA Satellites and Image Accuracy 1-E-1

    F. Characteristics of Water Vapor Imagery 1-F-1

    G. Satellite Interpretation 1-G-1

    H. Using Infrared Imagery to Monitor the Onset of Frost 1-H-1

    I. Ocean Thermal Features 1-I-1

CHAPTER 2

SYNOPTIC ANALYSIS 2-1

    A. Synoptic Scale Cloud Systems 2-A-1

    B. Cloud Patterns and the Upper Air Wind Field 2-B-1

    C. Air Flow Through Mid-Latitude Cyclones 2-C-1

    D. Satellite Interpretation 2-D-1

    E. Surface Cyclogenesis 2-E-1

    F. Locating Vorticity Centers 2-F-1

    G. Oceanic Cyclogenesis 2-G-1

    H. Extratropical Cyclogenesis over the Gulf of Mexico 2-H-1

CHAPTER 1

GENERAL INTERPRETATION

"Weather Service Forecasting Handbook No. 6 - Satellite Meteorology" is arranged into the following eight chapters:

1. GENERAL INTERPRETATION
2. SYNOPTIC ANALYSIS
3. PRECIPITATION
4. CONVECTION
5. TROPICAL WEATHER
6. FOG AND STRATUS
7. AEROSOLS
8. WIND AND TURBULENCE

The first chapter is designed to introduce the NOAA satellite meteorology program. This program includes the NOAA weather satellites, satellite data distribution, satellite imagery, and NOAA facilities that assist in imagery interpretation. The references in this chapter were selected to provide the background required to correctly interpret meteorological features on satellite imagery. The first five papers describe differences in imagery channels, thermal sensitivity in the infrared, cloud location errors, and atmospheric attenuation. The remaining four papers discuss basic introductions to such topics as water vapor imagery, cloud pattern identification, enhancement curves, and sea-surface temperature interpretation.

"The GOES User's Guide" should be mentioned at this point. The Guide is a non-technical description of NOAA geostationary (GOES) satellites, their data, and various patterns of data distribution. The User's Guide has not been included in the Handbook, because of its size and its general availability. Information in the following pages of the GOES User's Guide complements the information in this Handbook.

USER'S GUIDE SELECTIONS	MATERIAL COVERED
Sections 1 through 3:	GOES spacecraft and imaging systems.
Pages 4-1 through 4-6:	GOES data acquisition, distribution, interpretation and NOAA facilities responsible for these functions.
Pages 5-4 through 5-6:	Explanation of touchtone equipment used to select imagery sectors.

GENERAL INTERPRETATION

Pages 6-26 through 6-28: Description of the GOES imagery coded identification header.

Pages 7-6 through 7-11: Explanation of GOES infrared imagery enhancement.

Pages 7-12 through 7-51: Many examples of available enhanced imagery with a description of the purpose of each enhancement.

The next chapter, Synoptic Analysis, reviews basic principles of dynamic meteorology, and relates large-scale dynamic features and processes to features on imagery. The remaining six chapters deal with smaller scale or practical applications of satellite meteorology.

1-A Weather Satellite Interpretation - Introduction to Weather Satellite Imagery. David L. Carlson, NOAA Technical Memorandum NWS SR-103, National Oceanic and Atmospheric Administration, U.S. Department of Commerce, October 1982, 47 pp.

This reference provides a simple introduction to weather satellites, the NOAA data distribution system, and satellite imagery. Imagery characteristics and cloud and non-cloud features on imagery are briefly described. Succeeding references discuss the same material from an operational perspective.

1-B Reliability of Enhanced Infrared (EIR) Geostationary Satellite Data at High Latitudes. Frances C. Parmenter-Holt, Monthly Weather Review, 110 (10), October 1982, 1519-1523.

This reference compares observed surface temperatures to indicated surface temperatures on GOES imagery for areas that are near the image horizon (e.g., Alaska). Very low surface temperatures, comparable to storm-top temperatures, are discussed. Another reference in this chapter, Section 1-H, compares shelter temperatures with imagery-indicated surface temperatures closer to the GOES sub-point at warmer latitudes. These references provide information concerning the accuracy of the GOES infrared sensor.

1-C Displacement Error of Satellite Cloud Tops. Rich Warren, NWS Technical Attachment No. 77-G4, SSD-CRH, April 1977, 3pp.

Since the earth's surface is curved, a GOES satellite located over the equator views the United States at an angle. For this reason, tall cloud tops

GENERAL INTERPRETATION

on GOES imagery appear poleward of their true locations. This paper discusses the cloud displacement error, and provides a correction chart for the conterminous United States.

    1-D Cloud Location Corrections Near the Horizon of an SMS Image. Carl E. Weiss, Satellite Applications Information Note 78/8, NWS/NESDIS, U.S. Department of Commerce, Washington, D.C., 8pp.

This paper is the second of two papers describing cloud displacement errors. The previous paper discussed locations within the conterminous United States. This paper applies the same principles, and provides the same sort of correction chart to locations that are near the GOES image horizon (e.g., Alaska).

    1-E Supplementary Information from NESDIS Files. Satellite Applications Laboratory, National Oceanic and Atmospheric Administration, U.S. Department of Commerce, Washington, D.C., February 1986, 5pp.

Two selections from NESDIS files address topics not covered in the GOES User's Guide (see chapter introduction). The first selection introduces the NOAA polar-orbiting weather satellites. Polar orbiter imagery is available from SFSS GOES-TAP facilities and can be selected by using the Touchtone System.

The second selection summarizes factors to be considered and understood for successful interpretation of GOES imagery. Cloud displacement errors, a consequence of foreshortening, has been discussed in the previous two articles, 1-C and 1-D. The degree of atmospheric attenuation is described in reference 1-B.

    1-F Characteristics of Water Vapor Imagery. Roger Weldon and Sue Holmes, NESDIS Training Notes, Satellite Applications Laboratory, National Oceanic and Atmospheric Administration, U.S. Department of Commerce, Washington, D.C., October 1984, 2 pp.

These excerpts represent a succinct description of water vapor imagery, what the sensor measures, and factors to be considered in interpreting this imagery. The return from moisture imagery depends on the amount and vertical location of moisture and the temperature of the moisture bearing air column.

Relationships of moisture and temperature to imagery return are not simple. A return from low-level moisture, especially below the 700 mb level, will be warm and appear dry on the imagery. At high altitudes, the atmosphere is incapable

GENERAL INTERPRETATION

of holding enough moisture to produce a cold return. Moisture around the 400 mb level is most apt to show up on moisture imagery.

    1-G Satellite Interpretation. Eugene M. Weber and Steven Wilderotter, Third Weather Wing/Technical Note-81/001, Aerospace Sciences Division, Offutt AFB, NE, December 1981, 5-19.

This paper provides an introduction to the appearance of different clouds, and other features on imagery. Cloud patterns are defined, and examples of cloud types on imagery are given. Cloud patterns that indicate turbulence, wind speed and direction, and moisture advection are described and illustrated.

    1-H The Use of the HB IR Enhancement Curve to Monitor the Onset of Frost Conditions in the Western States. James J. Gurka and Ralph K. Anderson, Satellite Applications Information Note 76/25, NWS/NESS, U.S. Department of Commerce, Washington, D.C., 1976, 4pp.

This paper is included for its careful comparison of temperatures indicated by the GOES infrared sensor, and instrument shelter temperatures. Such factors as time, sensor resolution, and gridding errors are considered. Although the HB enhancement curve is no longer used, this paper addresses thermal sensitivity and other factors affecting infrared sensor temperature. Such a discussion is valid for any infrared imagery, since it describes the performance of a GOES sensor still in use.

    1-I Ocean Thermal Features as Seen from GOES-1. Stephen R. Baig, Satellite Applications Information Note 76/7, NWS/NESS, U.S. Department of Commerce, Washington, D.C., 1976, 4pp.

"Ocean Thermal Features" shows how the GOES infrared sensor is used to map water temperatures over large areas. Gulf Stream mapping is only one of many applications of ice and water temperature mapping, using GOES and NOAA polar orbiting satellites. This paper also describes the use of water temperatures to make inferences about low-level wind in data-sparse areas.

NOAA Technical Memorandum NWS SR-103

WEATHER SATELLITE INTERPRETATION - Introduction to Weather
Satellite Imagery

David L. Carlson
WSO/FAA Academy
Oklahoma City, Oklahoma

FAA ACADEMY
GOES SATELLITE COURSE

I. INTRODUCTION

Satellite imagery is an excellent source of observed data, and as such, it is a very helpful aid to the pilot briefer. Like other aids, such as surface observations and radar reports, it should never be used alone. A good briefer uses all available information to insure that pilots get the latest and most accurate weather picture possible.

One of the greatest advantages of satellite imagery is that the exact geographical coverage of a cloud system can often be seen. This is not possible with other types of observed data because of the large distances between observation stations.

GOES Satellite Photo Recognition

Figure 1 GOES West and GOES East Coverage.

CHAPTER 1
GOES IMAGERY

This lesson will introduce you to satellite imagery. It will briefly cover the satellite itself and the photo equipment on board, then it will cover in detail the types of imagery available from the satellite.

There are two main types of meteorological satellites; the Polar orbiters and the Equator orbiters. The Polar orbiters range in altitude from about 400 to 600 NM. They circle the earth in orbits that carry them over both poles. As the earth rotates below them, they can scan the entire globe, one strip at a time. Imagery from the Polar orbiters is routinely available in Alaska and Hawaii and is usually available on request in the remaining 48 states.

The GOES satellites are in orbits directly over the Equator. They are approximately 19,000 NM above the earth and make one revolution every 24 hours. Their speed is exactly timed with the rotation of the earth so they stay over the same spot on the earth all the time. The acronym GOES stands for *Geostationary* Operational Environmental Satellite. They may also be referred to as SMS, Synchronous Meteorological Satellite. Several GOES satellites launched by the U.S., Europe, and Japan provide global coverage plus backup capability. Two of these cover the U.S.; GOES East, located at 75 W. Long. and GOES West, at 135 W. See figure 1.

The scanning equipment aboard the Polar orbiter is very similar to that on the GOES satellite. The resultant imagery is also quite similar so photo interpretation is essentially the same for both types. This course will deal primarily with the GOES system.

The satellite contains two types of sensors, visible and infrared (IR). As the satellite spins at 100 rpm, these sensors scan the earth in horizontal lines starting at the North Pole and working down to the South Pole. When the sensors are facing away from the earth toward outer space, their aiming angle is lowered slightly. With each successive spin then, the sensors scan a horizontal line on earth directly below the preceding one. It requires 1,821 lines to make a full disc photo and at 100 rpm that takes 18.2 minutes. A new scan is started every 30 minutes. Approximately 30 minutes after the start of a new scan, a finished photo is available to users. That is *almost* real time information.

The visible sensor has eight identical channels arranged vertically as shown. Each time the scanner makes a horizontal sweep across the earth, eight lines of data are gathered, each line covering a strip ½ mile wide across the earth's surface. As a result, the full disc visible photo is composed of 14,568 lines of data (8 × 1,821 = 14,568) and the resolution can be as sharp as ½ NM, or in other words, anything ½ mile or more across is large enough to be reproduced on the photo. The IR sensor utilizes only one channel so the IR photo contains only 1,821 lines of data, giving it a resolution of 5 NM. Operationally, the visible photos you receive will have ½ NM, 1 NM, or 2 NM resolution, depending on the size of the area portrayed. GOES IR imagery will always have 5 NM resolution regardless of area size.

The raw data from the satellite are transmitted to ground stations in the Washington, D.C. area where it is processed through a number of mini-computers called sectorizers. A sectorizer extracts a portion of the data which corresponds to a certain geographical area. There are *standard* sectors and *floating* sectors. The sectorized data are then transmitted to the Satellite Field Service Stations (SFSS's) via telephone lines. The SFSS's distribute the data to the users also by telephone lines. (See figure 3.)

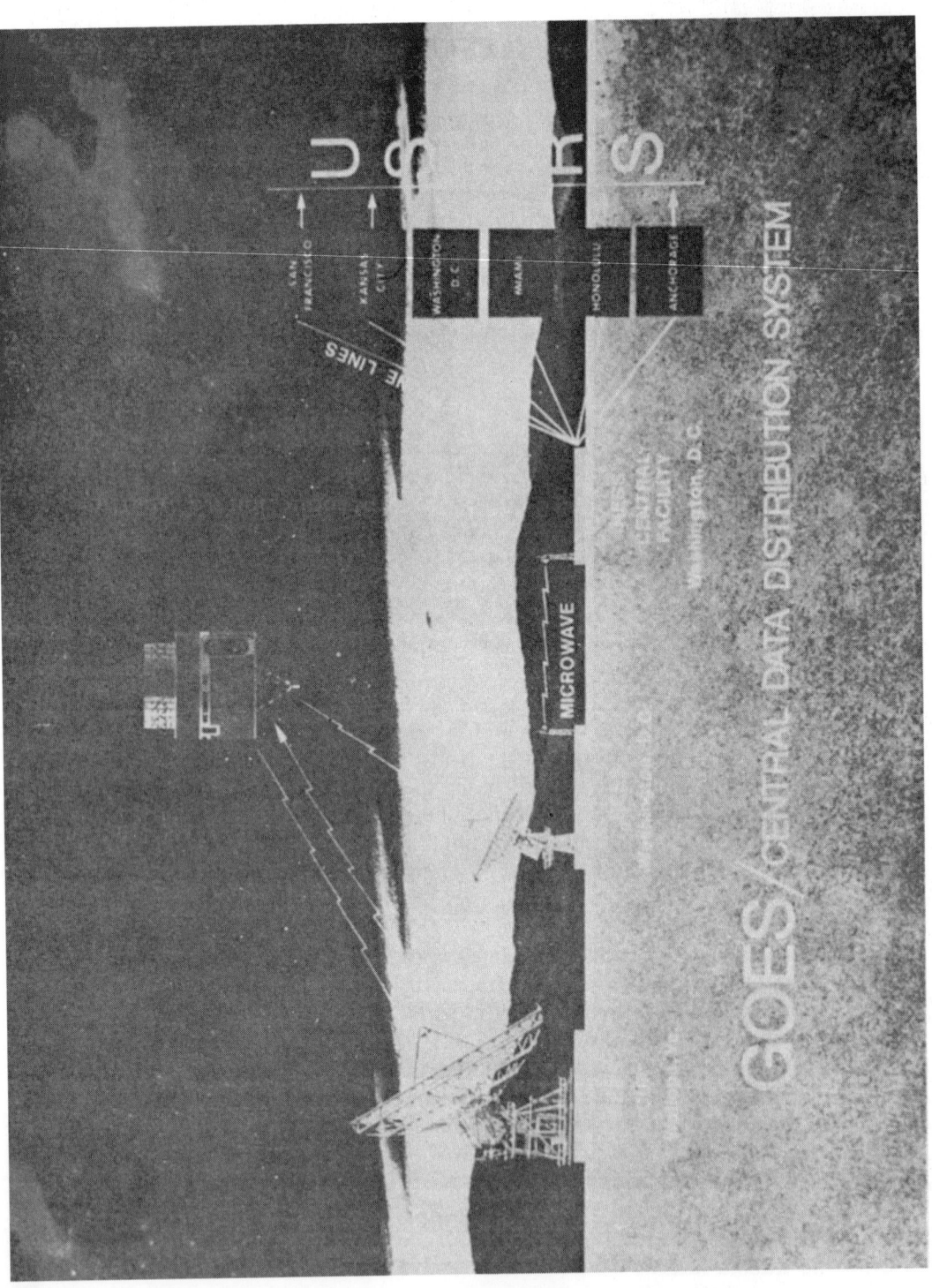

Figure 3 GOES/SMS Data Flow Diagram.

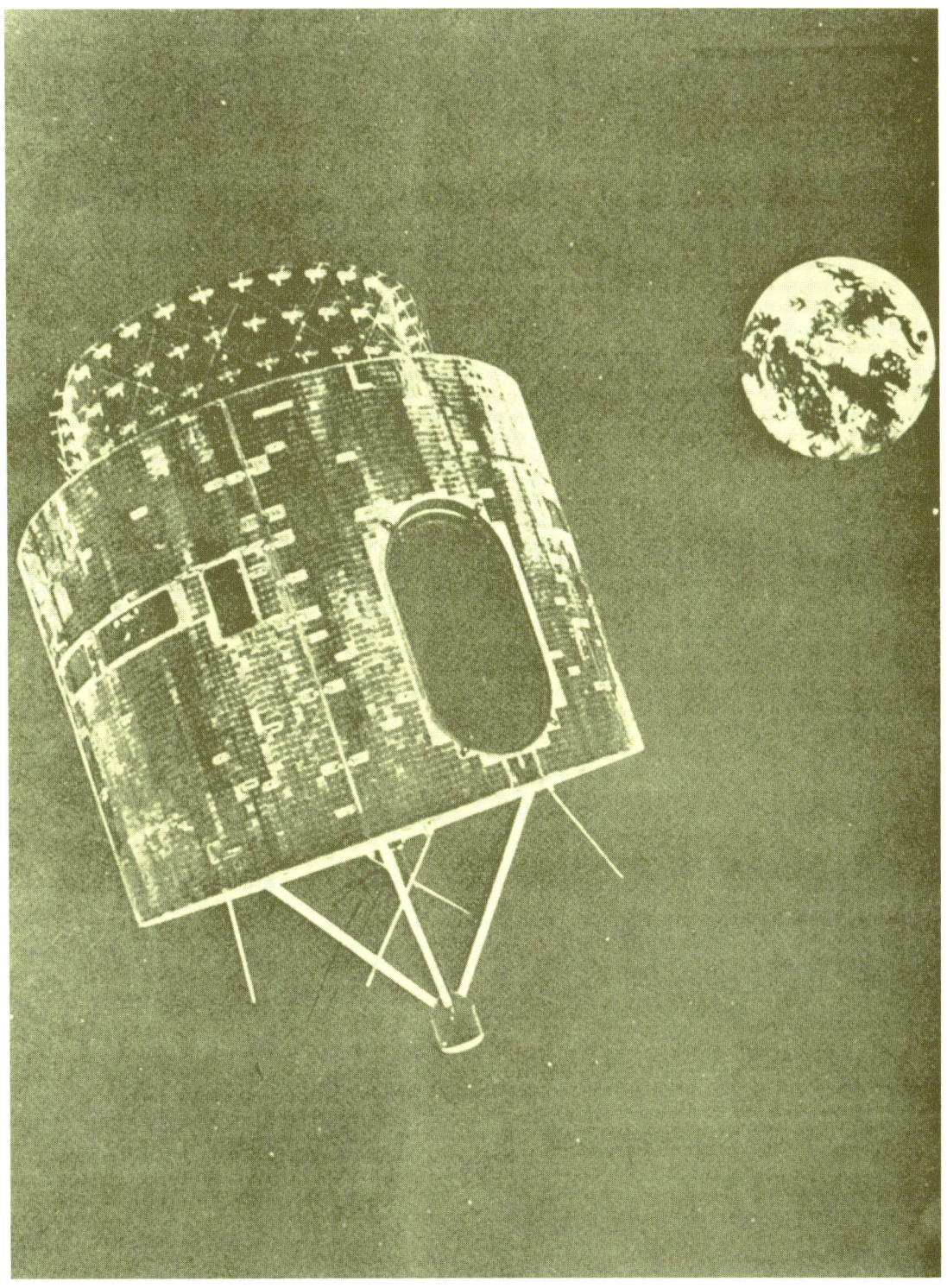

Figure 2 GOES/SMS.

GOES Satellite Photo Recognition

PHOTO NUMBER ONE

GOES Imagery

VISIBLE IMAGERY

This is a full disc visible photo from GOES East. (See photo No. 1.) The clouds are white, the land masses are gray, and the water areas are very dark, almost black. The visible photo is the result of reflected sunlight. Clouds are excellent reflectors so they appear very white. At night when there is no sunlight, there can be no visible imagery. Notice the black background. Neither the earth's atmosphere nor empty space reflect any sunlight, so any background area on a visible photo will always be black.

The reflectivity table (table 1) shows the percent of sunlight reflected by various surfaces. The best reflector of all is a large cumulonimbus, but all *thick* clouds reflect most of the sunlight that strikes them, so on a visible photo, thick clouds will all appear white. Thin clouds or areas of very small clouds will appear darker because much less sunlight is reflected.

Various types of terrain have intermediate or low reflectivity so land surfaces will appear as some shade of gray. Water surfaces are the poorest reflectors of all so they will appear almost black. Land-water contrast will normally be very good on visible imagery. Water will always be very dark unless it is very shallow, muddy, or frozen.

1. Large thunderstorm	92%	7. Thin stratus	42%
2. Fresh new snow	88%	8. Thin cirrostratus	32%
3. Thick cirrostratus	74%	9. Sand, no foliage	27%
4. Thick stratocumulus	68%	10. Sand and brushwood	17%
5. White Sands NM USA	60%	11. Coniferous forest	12%
6. Snow, 3-7 days old	59%	12. Water surfaces	9%

Table I   Reflectivity of various surfaces

Visible satellite photos look very much like standard black and white photographs. Actually, they are constructed here on earth, line by line, from digital information gathered by the satellite sensors, but they can be interpreted just as if they were snapped by a regular camera in space. In this respect, visible imagery can be thought of as an old familiar tool being used in a new unfamiliar area.

GOES Satellite Photo Recognition

PHOTO NUMBER TWO

1-A-8

GOES Imagery

GOES users seldom need a full disc photo; normally, you will want one of the sectors. This photo (No. 2) from GOES West portrays an area less than half the size of the full-disc photo. The resolution is 2 NM, so some of the small scale features cannot be seen. Look closely at the clouds over the ocean such as the area west and southwest of Baja, California (C). The clouds are not all uniformly white. Thin clouds or areas of small clouds show up as a gray shade. Again, this is because thin or small clouds just do not reflect as much sunlight as thick clouds.

Note the dark area in the upper left-hand portion of the photo. This is the sunrise-sunset line, commonly known as the *terminator*. That part of the world is still in darkness so it cannot be seen on visible imagery.

The fog and low stratus that is so persistent in the central valley of California in the winter can be seen at (D). The snow-capped Sierra Nevada mountain range (E) forms the eastern boundary of the valley. Large scale features such as these can usually be identified on visible imagery when they are not obscured by higher clouds. Identification of smaller scale features such as the Salton Sea in southern California (F) requires familiarity with the terrain and close inspection of the photo.

GOES PRODUCT LEGEND

A partial explanation of the GOES product legend is shown below. A complete explanation is available in the GOES User's Guide.

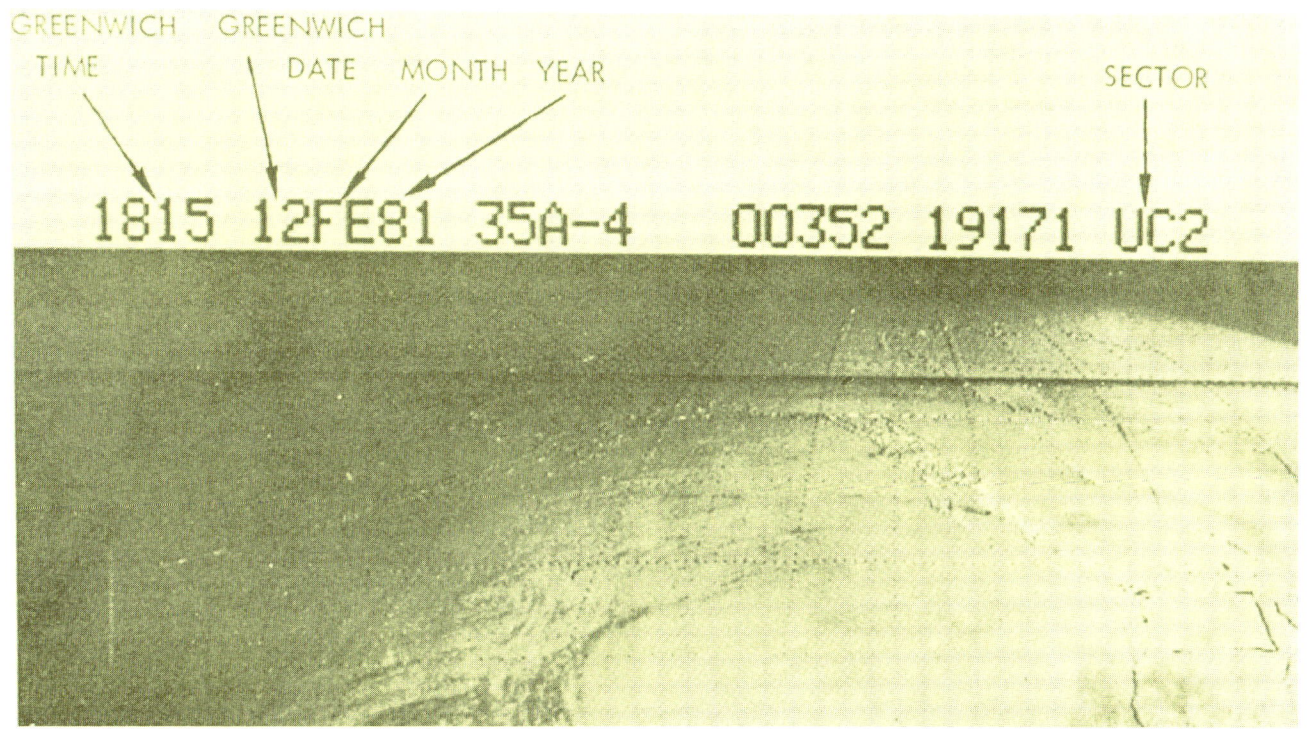

Figure 4.

GOES Satellite Photo Recognition
PHOTO NUMBER THREE

GOES Imagery

Photo 3 is from GOES East and shows a sector about half the size of the previous one. The resolution is 1 NM. Notice the extraordinary detail such as (A) the Continental Divide in Colorado and New Mexico. The heavily wooded mountain ranges show up much darker than the grassy valleys and rangeland nearby. The Black Hills area of South Dakota (B) is a very prominent landmark for the same reason. The White Sand Desert in south-central New Mexico (C) is another feature that can easily be seen, because the reflectivity of the white sand is much greater than that of the nearby darker soil.

The Great Lakes, the Gulf of Mexico, and various smaller lakes such as the Great Salt Lake in Utah show typical land-water contrast. The relatively thick clouds in the Ohio River Valley appear predominately white, whereas the thin, small clouds over central Nebraska and Iowa appear almost the same shade as the terrain.

There are other more subtle differences in shading that show much more detail than one would imagine. Look closely at the general area of eastern Texas, eastern Oklahoma, western Louisiana, and western Arkansas (D). It is a noticeably darker shade than most of the surrounding country. In the summer, this area is predominantly covered with green trees which do not reflect as much sunlight as the open rangeland to the west (E) or the open farmland in the broad, flat Mississippi River flood plain directly to the east (F). The flood plain of the Red River (G) is also easily seen from the northwest corner of Louisiana across the state toward the southeast where it empties into the Mississippi. The river itself is less than a mile across so it is too small to be shown on the photo, but the relatively treeless flood plain may be 10 to 15 miles across, and it shows up as a definite lighter shade than the tree-covered land on either side.

If possible, you should have a cloud-free photo for each season posted near your receiver for comparison purposes. This would be a great help to new people learning to use GOES products.

GOES Imagery

THIN OR SMALL CLOUDS

The subject of thin clouds or areas of small clouds deserves further clarification. Thin or small clouds, in themselves, are no hazard to aviation. A problem might arise, however, if a briefer were to give a pilot wrong information based on erroneous interpretation of a satellite photo. In that respect, it is important to be able to accurately interpret the shading on a photo, especially when there is some variation from what might be expected.

In an area of small or thin clouds (see fig. 5), part of the reflected sunlight sensed by the satellite is from the tops of the clouds and part is from the land or water surface below. The resultant gray shade on the photo, then, depicts an average of the two reflectivities. It is darker than a thick cloud area and lighter than the normal surface shade.

An exception to this would be if the surface were covered with snow or perhaps a thick layer of low clouds. Then there would be no noticeable error in shading. Under these circumstances, thin or small clouds may not be detectable at all.

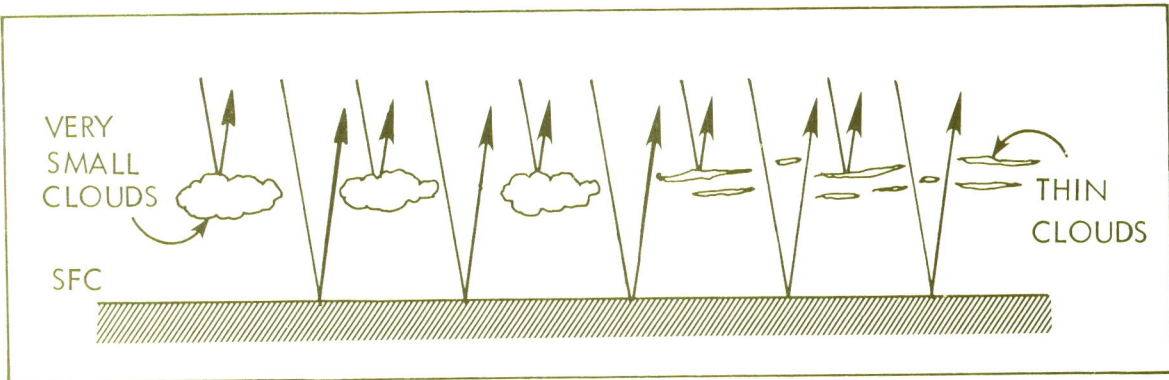

Figure 5.

To a small segment of the pilot population, thin clouds or areas of very small clouds are very important. Aerial photography and photogrammetry pilots usually need clear skies for their specialized missions. The low-altitude flights can often operate effectively in spite of thin middle or high cloud cover, but the high-altitude flights normally cannot tolerate clouds at any level regardless of how thin or small they are.

To meet the specialized needs of these pilots, the briefer must not only be able to accurately identify areas of thin and small clouds, but also be able to estimate their relative height. The subject of height determination will be covered later in the discussion of infra-red imagery.

GOES Satellite Photo Recognition
PHOTO NUMBER FOUR

GOES Imagery

SNOW COVER

Photo 4 is an example of the smallest sector size that is routinely available to GOES users. It covers an area less than half the size of the area covered in photo 3. The resolution is ½ NM. so much more geographical detail can be seen than on the larger sectors.

Snow cover is often very difficult to identify on satellite imagery. Like clouds, snow reflects most of the sunlight that strikes it (see table I, pg 7), consequently, clouds and snow cover may look exactly alike, especially over relatively flat terrain. The most reliable indicator for differentiating between clouds and snow cover is being able to recognize known terrain features such as unfrozen rivers and large lakes. Clouds normally obscure terrain features, but snow cover does not.

In photo 4, the states of Nebraska, South Dakota, and Wyoming are nearly cloud-free and mostly covered with snow. Note the Missouri river in South Dakota (A). It is clearly visible in central and southeastern South Dakota and can be identified as far south as Omaha, Nebraska (B). The Platte River (C) is identifiable from Omaha westward almost to the Wyoming border. In fact, someone who is very familiar with the topography of Nebraska could identify most of its major rivers and several of its large lakes on this photo. To someone unfamiliar with this area, these dark lines and spots may look like shadows on a low cloud layer or perhaps holes or thin spots in a cloud layer.

Snow in mountainous country is usually easier to identify because it often forms a dendritic (branchy) pattern. Mountain ridges above the tree line are essentially barren, and the snow is visible there, but in the tree-filled valleys, most of the snow is hidden beneath the trees (see fig. 6). Two good examples are shown at D and E on the photo. Mountain areas such as the Black Hills (F) that are completely covered with trees do not exhibit the dendritic pattern but are still easily recognized by the sharp contrast with the nearby open country.

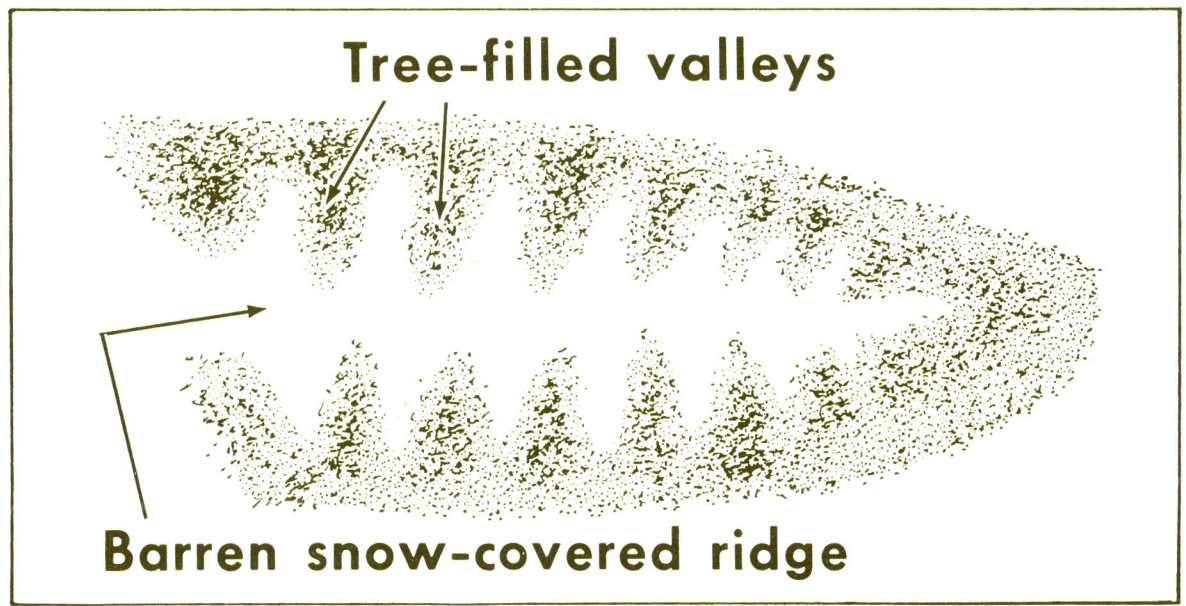

Figure 6.

There are several sources of observed data to help you determine snow cover. Surface observations and pilot reports are very helpful, especially when they confirm that the sky is clear. Also, you can compare current photos with earlier photos. Cloud patterns can change noticeably in a few hours, whereas snow cover normally changes very little from day to day.

Finally, if you have access to meteorological facsimile charts, the National Weather Service issues an Observed Snow Cover chart once a day during the winter season.

GOES Satellite Photo Recognition

PHOTO NUMBER FIVE

GOES Imagery

INFRA-RED IMAGERY

This is a full disc IR photo. It is from GOES East and was taken at the same time as photo No. 1. IR imagery is just a picture portraying different temperatures as black, white, or some shade of gray.

Everything with a temperature above absolute zero radiates electromagnetic energy. The wavelength of this radiation varies with the temperature. As energy radiates from the surface of the earth and the tops of clouds, the IR sensor measures the energy level at specific wavelengths. The energy measurements are transmitted to the receiving station in Washington, D.C. where they are converted to temperatures. The temperatures can then be shown as black, white, or some specific shade of gray. High clouds are very cold so, on this photo; they appear white; for example, the clouds near (A). Mid-level clouds are somewhat warmer so they are a light gray shade (B). Low clouds are warmer still, so they are a darker shade of gray (C). Often, low clouds are the same temperature as the surrounding terrain and cannot be distinguished at all. In that case, you need a visible photo or surface observations to detect them. The low clouds at C are easily seen on photo 1 but are not easily seen on photo 5.

Land-water contrast is also dependent on temperature contrast. In this photo the coast of California is easily seen because the land is warmer, so it is noticeably darker. At night, however, the land cools rapidly and may become cooler than the water. Then it would appear as a lighter shade. If the water and the land are approximately the same temperature, such as around the Yucatan Peninsula in Central America (D), they will also be the same shade of gray, and there will be no land-water contrast.

The graph in Figure 7 shows exactly how the unenhanced IR photo is shaded. It is simply a straight-line relationship between temperature and gray shades. Temperature is shown on the bottom along the horizontal axis, and shades between black and white are on the vertical axis. The temperature range covered is from a plus 134°F (56.8°C), which might be found in the Sahara Desert in summer, to a minus 165°F (−109°C) which might be found at the tops of very high clouds (about 60,000 ft.). The graph shows that 134°F (56.8°C) would be black, a minus 165°F (−109°C) would be white, and every temperature in between would be some specific shade of gray.

The computer is capable of producing and recognizing 256 distinct shades from black to white. The human eye, however, can only distinguish from 15 to 20 shades, depending on conditions; so the unenhanced IR photo is not used much operationally. A technique called Enhancement is used to highlight areas of interest.

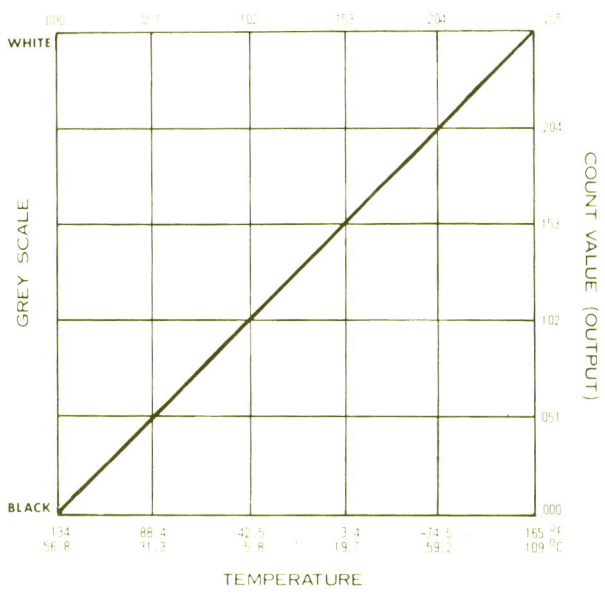

Figure 7.

1-A-17

GOES Satellite Photo Recognition

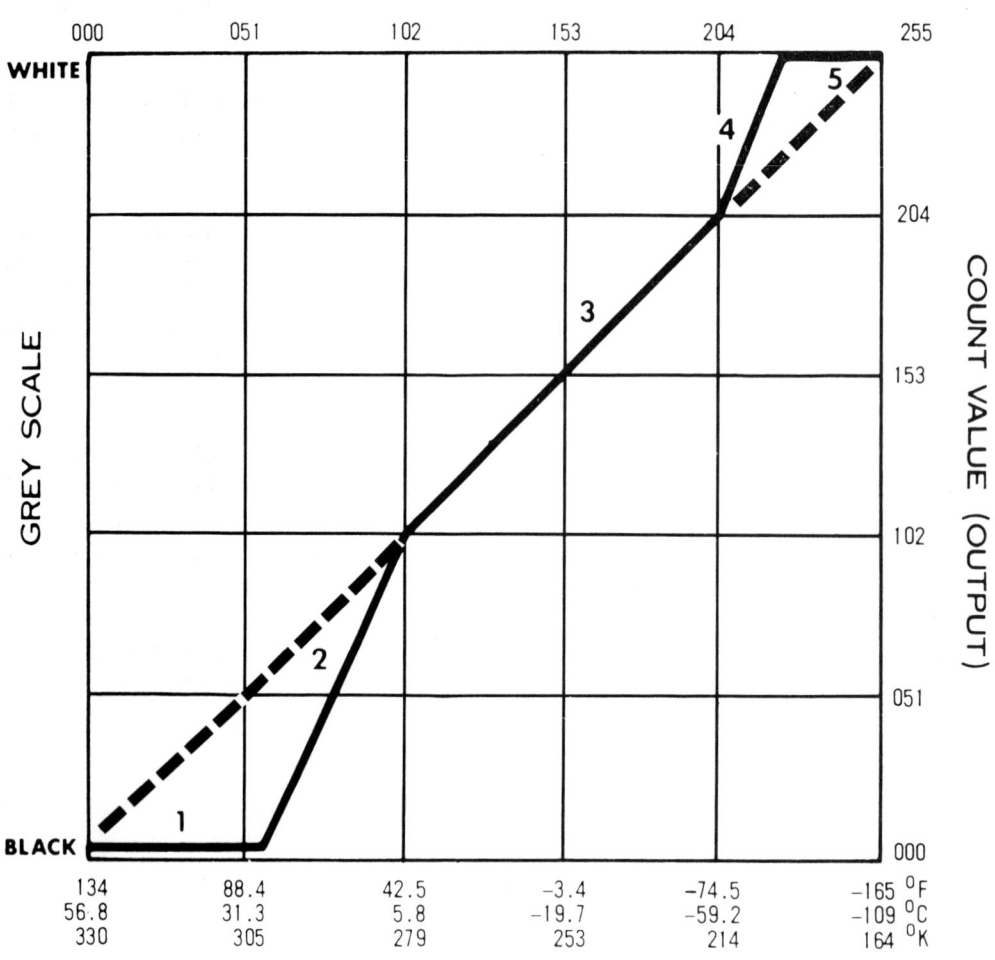

Figure 8.

GOES Imagery

ENHANCEMENT

On the unenhanced IR photo, each degree of temperature was represented by a slightly different shade of gray. In the enhancement process, any shade of gray may be assigned to any temperature when more contrast is needed to highlight a certain temperature range.

The ZA curve (see figure 8, solid line) is a slight modification of the unenhanced curve (dashed line). The very warm temperatures in segment 1 (56.8°C to 29.3°C) are all shown as black. Since there are no clouds in that temperature range, there is no need for a difference in shading.

Segment 2 (28.8°C to 6.8°C) contains the temperatures of most low clouds and sea surfaces. In this way, the darker shades of gray are used over a smaller temperature range so that small differences on temperature can be more easily detected.

A similar process is done at the other end of the temperature scale except that it affects a much smaller range of temperatures. All temperatures of minus 75.2°C or colder are shown as pure white (segment 5). Here, too, since there are so few clouds in this temperature range, there is no need for a difference in shading. The very light gray shades out to white are used in the temperature range of the very high cirrus clouds (segment 4, $-56.2°C$ to $-75.2°C$).

Between these two extremes in segment 3 (6.3°C to $-55.2°C$), no enhancement is used. Clouds in that temperature range will be shaded the same as on the unenhanced IR photo. The ZA curve can be thought of as a somewhat improved version of the unenhanced curve, and the ZA infrared photos are generally used in place of the unenhanced infrared photos. The end result of the ZA curve is that the temperature range we are observing has been narrowed somewhat by effectively eliminating the very warm and the very cold temperatures.

Interpretation of the ZA photo is essentially the same as the unenhanced IR photo because *in the temperature range of interest*, each degree of temperature has its own separate shade of gray which is not used for any other temperature.

GOES Satellite Photo Recognition
PHOTO NUMBER SIX

GOES Imagery

THE ZA PHOTO

This is an example of an IR photo using the ZA enhancement curve. Operationally, the ZA curve is used quite often, especially at night. This photo is approximately the same time as photo #2. Note the excellent land-water contrast along the coasts of Mexico. Note also, there is good contrast between the sea surface and the lower clouds in warm southern Pacific (A) but poor contrast in the cold northern Pacific (B). The low clouds in that area are almost the same temperature as the sea surface.

The fog and low stratus in the central valley of California (D) can be seen on this photo as well as photo 2. The temperature at the top of the stratus layer is about the same as the temperature of the higher terrain surrounding the valley; however, the sun has heated the valley floor around the edges of the fog and the warm surface temperatures appear as a dark outline around the fog area. At night the warm surface will cool rapidly so that by early morning the entire area will probably appear as one light shade of gray. Temperature contrasts at the surface and between the surface and low clouds are usually at a minimum just when they are needed most, in the early morning before visible imagery is available.

GRAY SCALE WITH TEMPERATURE VALUES

Every enhanced IR photo has a gray scale directly below the product legend. It is a visual explanation of the enhancement curve used (see Fig. 9 and photo 6). The vertical lines are 10-degree increments of temperature in degrees Celsius. This temperature scale is shaded the same way the photo is shaded so you can tell at a glance what temperature or range of temperatures is depicted by any shade on the photo. Notice that to the right of $-30°C$ the vertical lines are only half as far apart. They still represent 10 degrees of temperature change, however.

Notice, also, that in the legend, just to the right of the enhancement curve identifier, there is a 5-digit group beginning with zero. The vertical line representing $0°C$ will always be just below the first zero of that group.

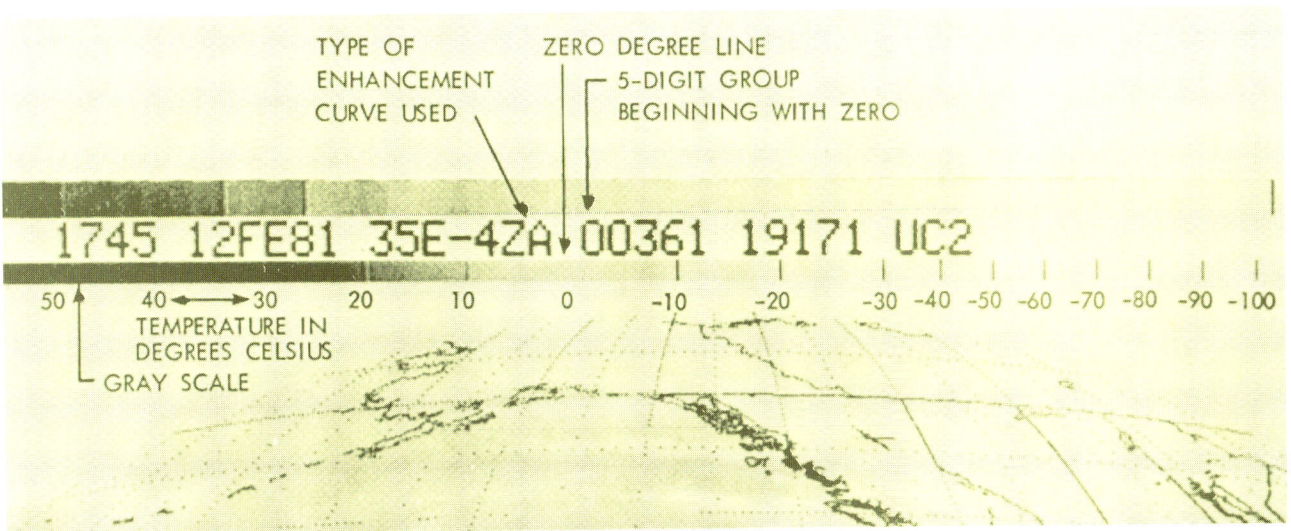

Figure 9.

THIN OR SMALL CLOUDS ON IR IMAGERY

Thin or small clouds create errors in shading on IR photos as well as visible. Assume an area is exactly half-covered with small clouds (see Fig. 10). The tops of the clouds are at 10,000 ft., and the temperature at the tops of the clouds is 0°C. The temperature of the surface at sea level is 20°C. Half of the radiation coming from that area would be from the tops of the clouds and half from the earth's surface. The satellite would sense an average temperature of 10°C, and the resultant gray shade on the photo would correspond to 10°C.

Again, thin or small clouds are not a hazard to aviation but GOES users must be aware of the shading errors they produce.

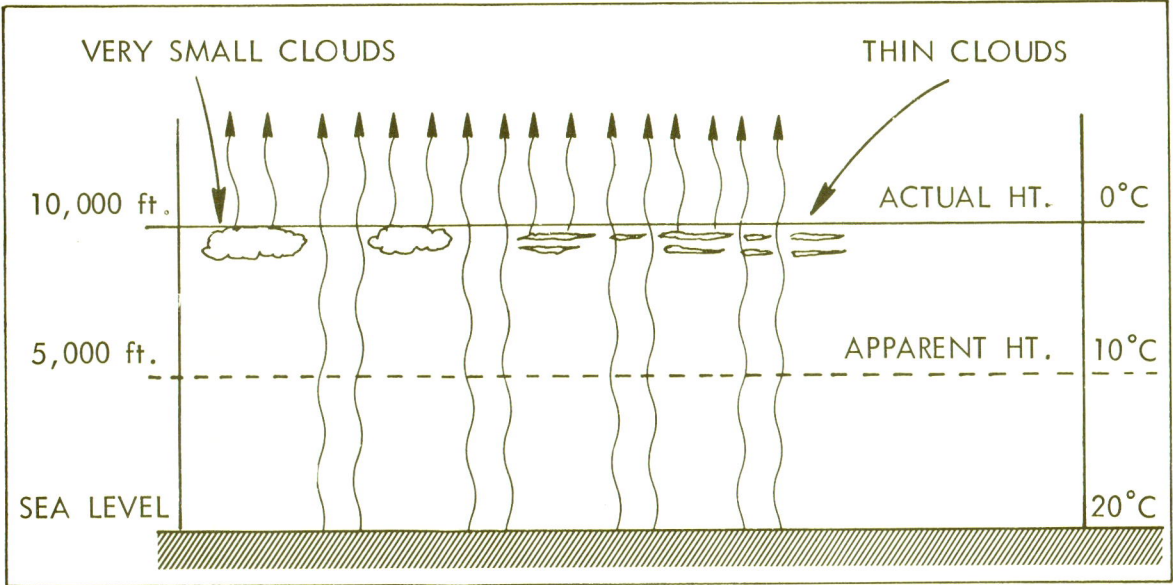

Figure 10.

GOES Satellite Photo Recognition

PHOTO NUMBER SEVEN

ENHANCED IR COMPARISON 2130Z 26 Mar. 76

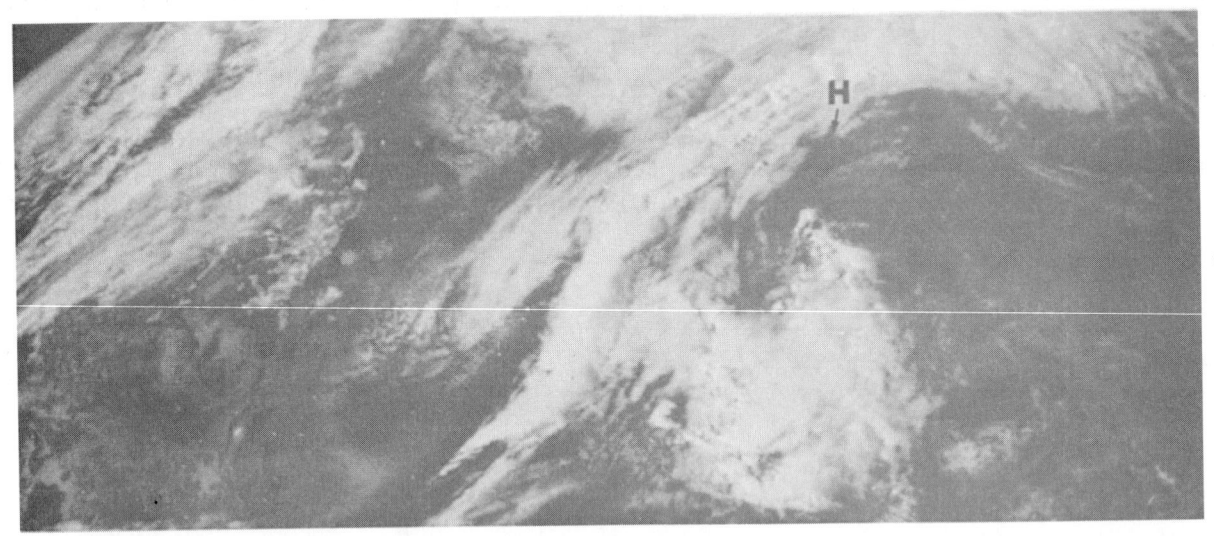

VIS

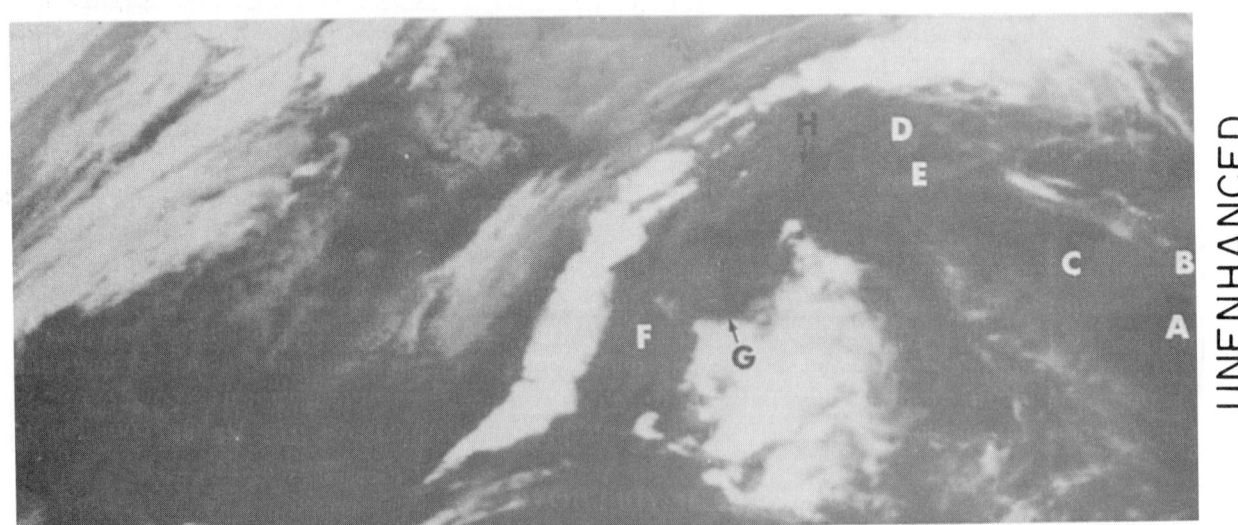

UNENHANCED

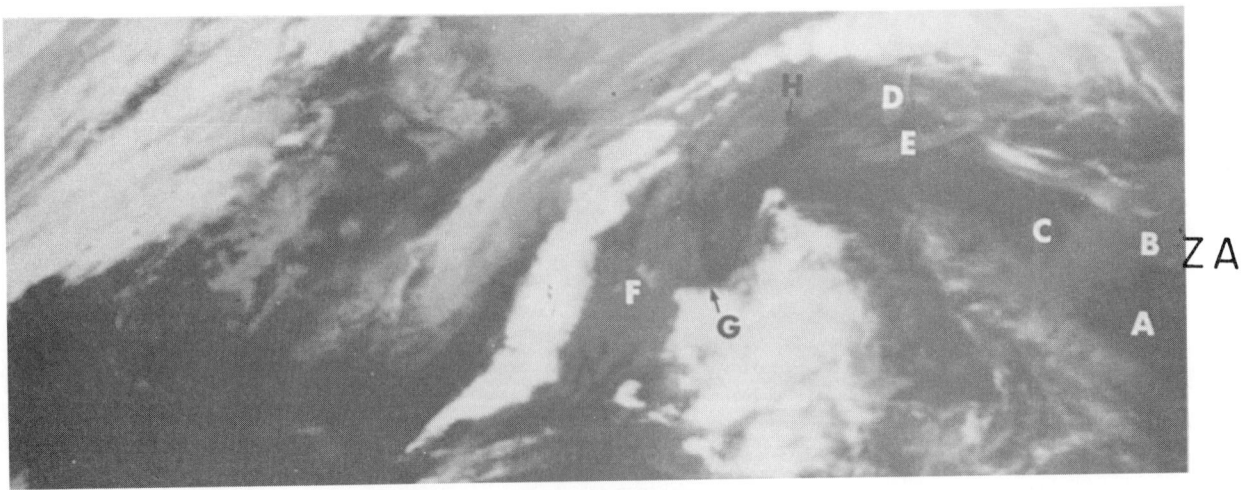

GOES Imagery

INTERIM SUMMARY

Photo 7 is a comparison of the three types of photos we have discussed so far. On the visible photo, the clouds are essentially all white. Only areas of thin clouds or small clouds appear as a somewhat darker shade. Terrain varies from medium gray to dark gray, depending on type of soil and foliage. Water areas are almost black.

On the unenhanced IR photo, each degree of temperature is shown by some shade of gray between black and white. Very cold temperatures are shown as white and very warm as black. A rough comparison of cloud height can be made from the different shades. Keep in mind that cloud families are classified according to the height of their bases, but the satellite sees only tops. Therefore, we can only compare relative heights of cloud tops.

The ZA curve photo is very similar to the unenhanced photo. The major difference is more contrast in the warmer temperatures of the land surfaces, water surfaces, and low clouds on the ZA photo. Note the more pronounced difference in the shading between the warm waters of the Gulf Stream (A), the cold waters of the Labrador Current (B), and the East Coast of the U.S. in the Delaware Bay and Chesapeake Bay area (C). The cool waters of Lake Huron (D) and Lake Erie (E) are other examples of better land-water contrast. Also note the increased contrast between the low clouds at (F) and the nearby cloud-free area (G).

The visible photo shows that Lake Michigan (H) is partially covered by clouds. Both of the infrared photos show that the water temperature and the cloud top temperature in that area are about the same. Because these temperatures are the same, neither infrared photo can show how much of the lake is covered by clouds and how much is not. Without the visible photo, the location of the edge of the clouds over the water cannot be seen.

During daylight hours, when both visible and infrared imagery are produced, the maximum amount of satellite information is available. The location of the clouds can be determined from the visible photo and their relative heights (high, middle, or low) can be estimated from the infrared photo. During the hours of darkness, when only infrared imagery is produced, high level and middle level clouds are usually readily apparent, but the warmer low clouds may not be distinguishable.

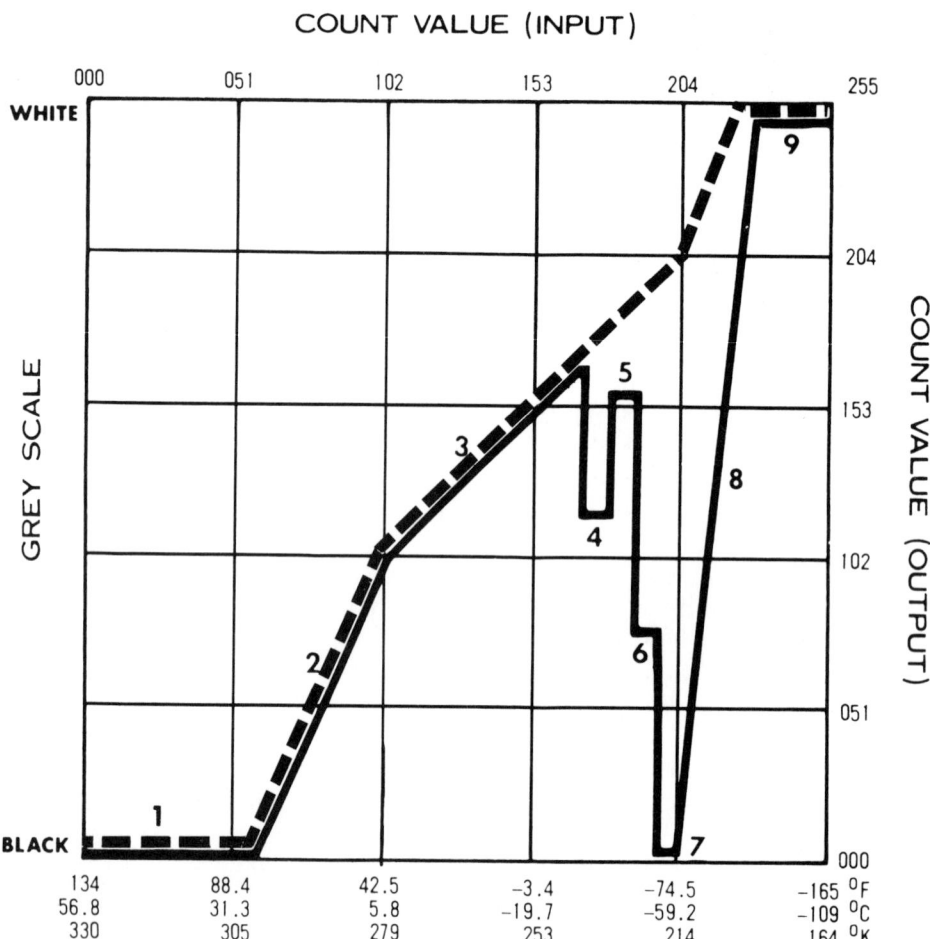

FIGURE 11.

THE MB CURVE

The enhancement curve that is used most is the MB curve. The solid line in Fig. 11 is a graph of temperature versus shading of this curve. The dashed line represents the ZA curve. Note that the warm part of the MB curve is the same as the ZA curve (segments 1, 2, and part of 3). The MB curve is meant to show overshooting tops of thunderstorms so the real enhancement starts at the cirrus cloud temperature level ($-32.2°C$). See Fig. 12.

GOES Imagery

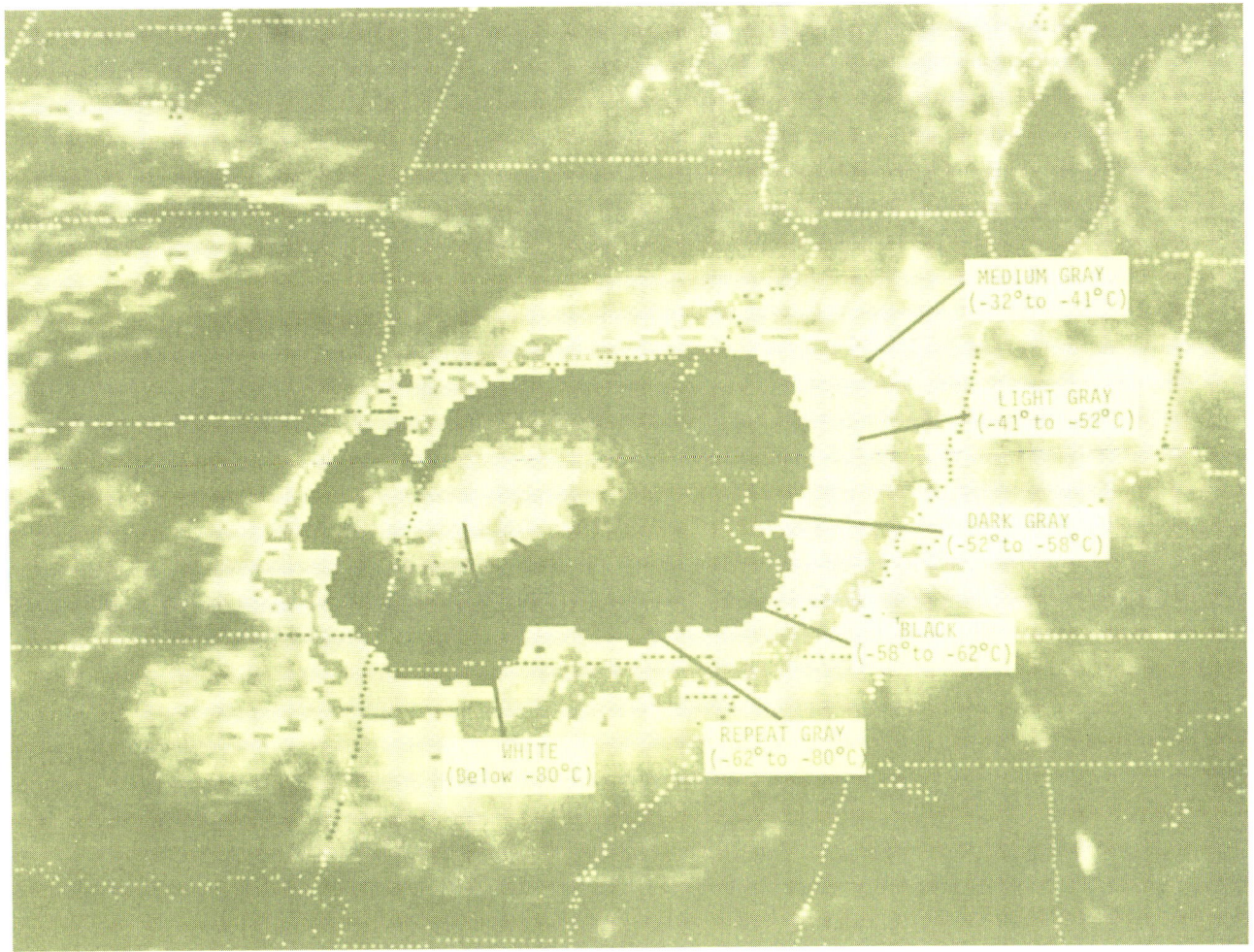

Figure 12.

Figure 12 is a magnified, detailed illustration of segments 4 through 9 of the MB curve. Temperatures are rounded off to whole numbers for convenience. The process of using contrasting gray shades for progressively colder temperature ranges forms a contour pattern. (Refer to Fig. 11 as well as Fig. 12 as you proceed.) Segment 4 (-32 to $-41°C$) forms the first contour. It is shaded medium gray to make it stand out plainly. Segment 5 (-41 to $-52°C$) is shaded light gray, segment 6 (-52 to $-58°C$) is dark gray, and segment 7 (-58 to $-62°C$) is black. In this way, the areas of intense up-drafts are clearly defined because the clouds formed in these up-drafts are higher and, therefore, colder than the surrounding cirrus clouds.

Segments 8 and 9 utilize all the shades from black to white and indicate cloudtop temperatures colder than minus 62°C.

The MB curve was designed primarily to show areas of convection. It must be added that high clouds with very cold tops do exist in nonconvective areas and in areas where convective activity is dissipating or has ceased. The MB curve will enhance these clouds also. A contoured area on an MB photo with at least a black contour (segment 7) is *usually* a good indicator that convective activity is occurring, but other sources of information should be used to verify the actual presence of thunderstorms and to determine their severity.

GOES Satellite Photo Recognition

PHOTO NUMBER EIGHT

GOES Imagery

THE MB PHOTO

This is an MB photo at midday in July. Most of the Central U.S. is shaded black. This is quite common in the summer. Refer to the gray scale on the photo and to Fig. 11. The very warm temperatures, 56.8°C to 28.2°C (134°F to 83°F) are all black. In other seasons, especially winter, surface temperatures do not get that high, so the only black areas would be the enhancement in the very cold (-59.2°C to -62.2°C) high cloud tops. Also, surface temperatures usually cool to below 83°F at night even in summer, so an early morning IR photo will be predominately gray and may appear quite different from the daytime photos.

This photo shows what is most likely a large cluster of thunderstorms covering most of eastern Kentucky. The contour pattern indicates a large area of very high tops with temperatures colder than minus 62°C (see Fig. 12). Now look at these same clouds on photo 3. They look essentially the same as nearby lower clouds. The visible photo by itself gives no reason to suspect possible severe weather in that area.

Refer to photo 3 and look at the thin clouds in Nebraska and South Dakota (H) extending eastward through Iowa (I) to the Lake Michigan area. Then look at those same clouds on photo 8. Notice the marked difference in shading on photo 8 from H to I. The thin clouds near H are high clouds, and their cold temperatures make them easier to see than the lower, warmer thin clouds to the east (I). Both areas are subject to the shading errors that thin clouds produce, but thin high clouds can usually be recognized on an IR photo; whereas, thin low clouds are often undetectable without a visible photo for comparison.

For each of the first three visible photos (1, 2, and 3) there is a corresponding size in infrared imagery (5, 6, and 8). Both enhanced and unenhanced infrared imagery are available in these three sizes. (The resolution of GOES infrared imagery is 5 NM. regardless of sector size.)

For the smallest sector size, such as photo 4, however, the equivalent infrared size can be obtained but is not always available. To compare this size visible photo with infrared imagery, you will often have to use a larger sector size infrared photo such as the size of photo 8. When comparing two photos of different sector size, keep in mind the difference in size of the area portrayed.

SNOW COVER ON IR

In identification of snow cover, the unenhanced infrared and the two most common infrared enhancement curves (ZA and MB) are usually no help at all. Normally, there is not enough temperature contrast between the snow cover and the adjacent surface to appear as contrasting shades of gray. On days when the sky is clear, barren terrain that is not covered by snow *can* become significantly warmer than nearby snow-covered terrain, at least for a few hours in the afternoon. Under these conditions, the boundaries of snow cover can be identified on infrared imagery. However, under these ideal conditions, snow cover can be seen in detail on visible imagery.

There are enhancement curves developed specifically to show snow cover and snow melt information, but they are used mostly for hydrology rather than pilot briefing and are seldom requested by aviation users.

GOES Satellite Photo Recognition

PHOTO NUMBER NINE

ENHANCED IR COMPARISON 2130Z 26 Mar. 76

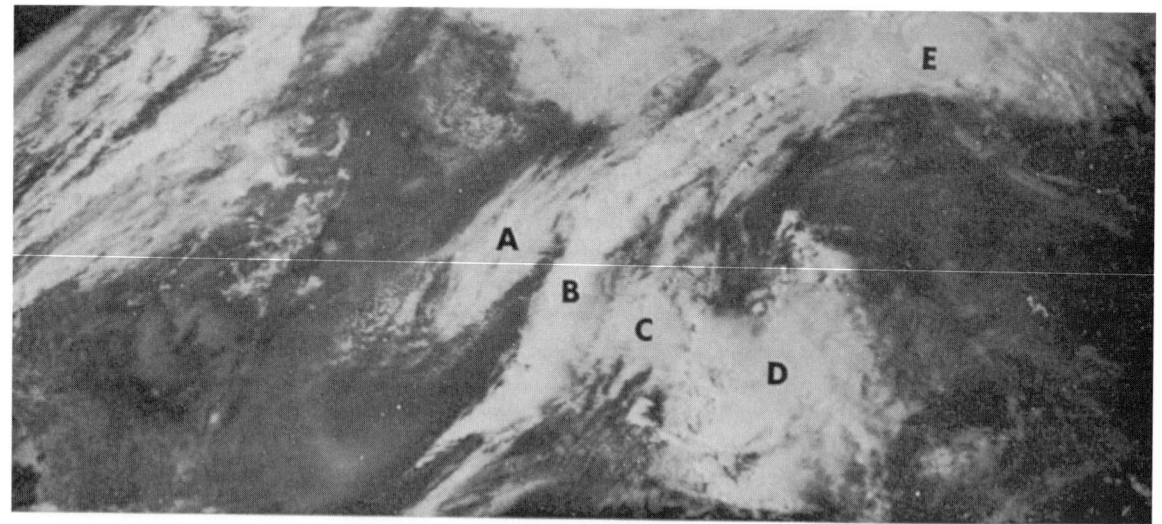

VIS

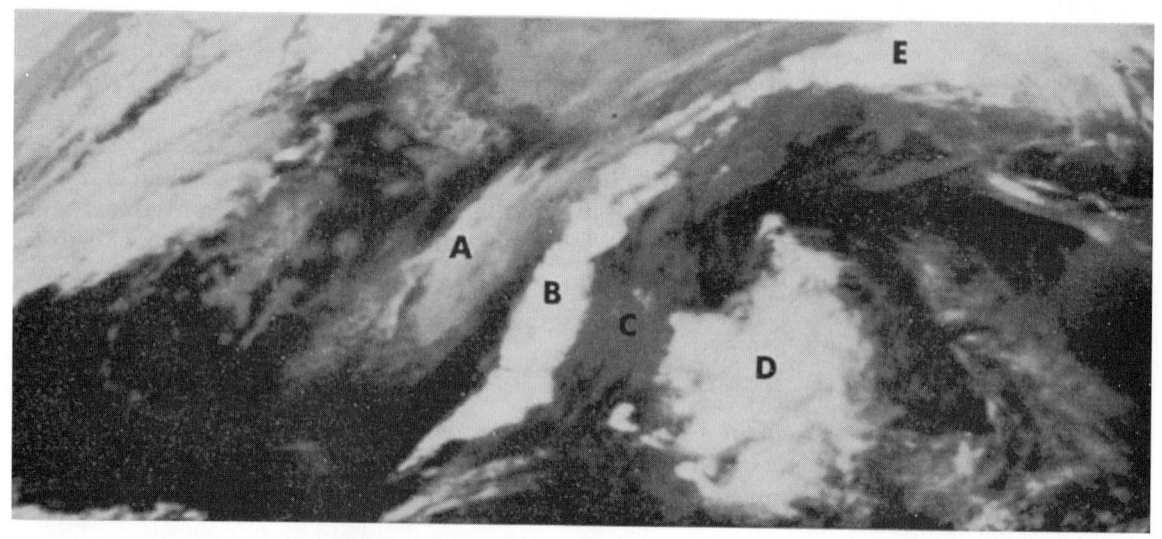

ZA

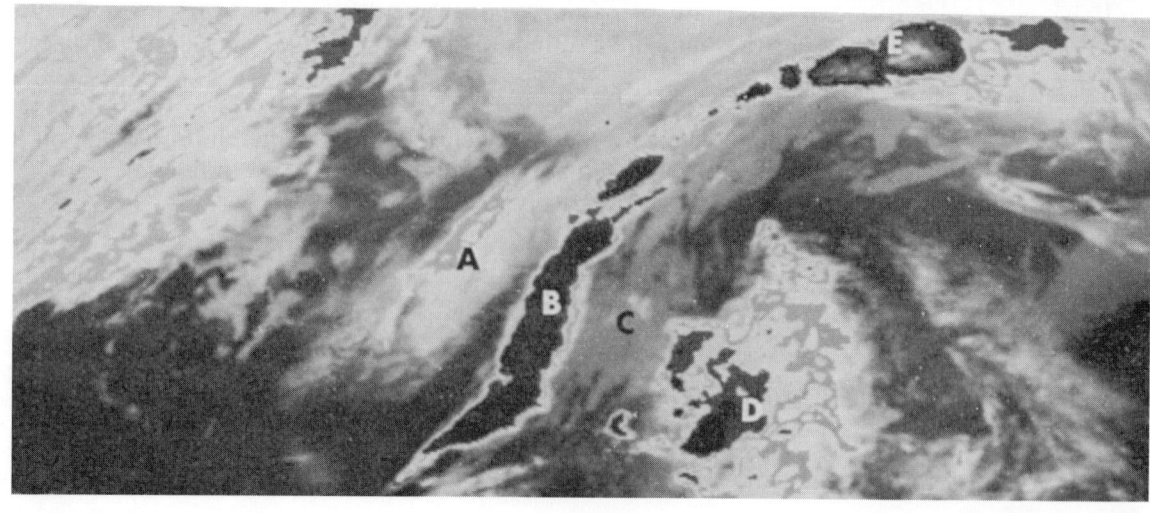

MB

1-A-30

GOES Imagery

SUMMARY

Photo 9 is a comparison of the three types of photos you will see most often in day-to-day use.

On the visible photo, all thick clouds will be essentially the same shade, almost white. Thin or small clouds will appear darker than thick clouds because they do not reflect as much sunlight. Differences in shading do not provide any information about cloud height, only cloud thickness. Thick clouds may look very much alike regardless of their altitude. (Refer to A, B, C, & D on the visible and ZA photos.)

The ZA curve photo is very similar to the unenhanced IR photo except that the very warm temperatures are all black and the very cold temperatures are all white. In between, from 28.8°C to −75.2°C (83°F to − 103°F) each degree of temperature has its own shade of gray which is not used for any other temperature.

The MB curve is similar to the ZA curve in the warm temperature range, from very warm surface temperatures up through the mid-level cloud temperature range. In the cold high cloud temperature range, contrasting gray shades are used to highlight the high, cold overshooting tops of thunderstorms (B, D, & E).

Note on the ZA photo, there is a large area of clouds in the vicinity of D that is shaded white. This indicates that the tops of all these clouds are high level but it does not give any indication that some of the tops are higher than others. The MB photo shows that only a small portion of this area contains overshooting tops of thunderstorms. The majority of the area consists of a thick cirrus layer which is located at a somewhat lower altitude. Also, it is evident that the coldest (and therefore probably the highest) clouds on the photo are located at E.

Look closely at other areas on the visible photo and note how each area appears on the ZA and the MB photos. Keep in mind that all infrared photos are nothing more than representations of temperatures. Surface temperatures can vary greatly from day to night so the shading on infrared photos can also vary considerably in a 24-hour period.

The ZA and the MB are commonly used enhancement curves, but there are many others. Over a hundred have been evaluated by NESS and about a dozen are available to you at any one time from your Satellite Field Service Station (SFSS). It is beyond the scope of this text to illustrate a large variety. If you see a photo utilizing a curve that is new to you, study the shading strip very carefully to see exactly what temperature range or ranges have been enhanced. This will give you a good idea of what features are being highlighted and will help you interpret the shading. If you want to discuss a certain enhancement curve or have a question about photo interpretation, call your SFSS. They are on duty 24 hours a day, and providing guidance to users of GOES products is an important part of their job.

CRITERION TEST
Chapter I

1. The visible photo is a result of

    a. reflected sunlight
    b. terrestrial radiation
    c. absorbed solar radiation

2. The brightest feature on visible imagery is

    a. fresh, new snow
    b. a frozen lake surface
    c. a large thunderstorm

3. On visible imagery, water will appear darker than land unless the water is

    a. muddy
    b. frozen
    c. either muddy or frozen

4. Infrared imagery is the result of

    a. reflected sunlight
    b. terrestrial radiation
    c. reflected solar radiation

5. On unenhanced infrared imagery, water will appear darker than land if the water is

    a. muddy
    b. warmer
    c. colder

6. Thin clouds will appear darker than thick clouds on

    a. visible imagery
    b. unenhanced infrared imagery
    c. both visible and unenhanced infrared imagery

7. Enhancement is used in infrared imagery to

    a. increase the resolution
    b. highlight areas of interest
    c. amplify surface reflectivity

GOES Satellite Photo Recognition

8. The smallest resolution on GOES infrared imagery is

    a. ½ NM.
    b. ½ KM.
    c. 5 NM.

9. What would be the darkest feature on infrared imagery that has been enhanced using the ZA curve?

    a. a hot land surface
    b. an ocean surface
    c. the overshooting top of a thunderstorm

10. On a visible photo, the darkest feature would be

    a. a hot land surface
    b. an ocean surface
    c. the overshooting top of a thunderstorm

11. What would be the darkest feature on infrared imagery enhanced using the MB curve?

    a. a hot land surface
    b. the overshooting top of a thunderstorm
    c. both a and b

12. Which surface would appear darker on visible imagery?

    a. a thick forest
    b. open rangeland
    c. a desert

13. The ZA curve and the MB curve are

    a. completely different
    b. alike in the cold temperature range
    c. alike in the warm temperature range

14. Snow cover is difficult to identify by the use of visible imagery alone because it looks very much like

    a. cloud cover
    b. barren terrain
    c. tree-covered terrain

15. What type of satellite imagery is most useful as an aid in identification of snow cover?

    a. Enhanced infrared
    b. Unenhanced infrared
    c. Visible

16. Select the most accurate statement concerning snow cover and cloud cover on visible imagery.

    a. Snow cover obscures prominent terrain features but cloud cover does not.
    b. Cloud cover obscures prominent terrain features but snow cover does not.
    c. Both cloud and snow cover obscure prominent terrain features.

Chapter II

CLOUD TYPES ON SATELLITE IMAGERY

Clouds are formed when air is cooled to its dew-point. The most common way that nature cools air is by lifting it. As air rises, it cools and clouds are formed. Most of the clouds you see on satellite imagery are the result of one or more of nature's lifting mechanisms.

Another way that air can be cooled is by contact with a cold surface. When this happens, the cloud forms on the ground and we call it fog. On satellite imagery, fog is just another low cloud.

Basically, there are only two types of clouds: cumuliform and stratiform. Either type may form at any level: high, middle, or low.

Cumuliform clouds are rounded, billowy, and puffy because they are formed in unstable air. Stratiform clouds are flat and sheetlike because they are formed in stable air. As such, clouds are excellent indicators of atmospheric stability.

Aviation hazards associated with stable conditions are quite different from those associated with unstable conditions. *In general,* unstable conditions are associated with thunderstorms, showery precipitation, gusty winds, low level wind shear, convective turbulence, clear icing, good visibility, and an absence of low ceilings. Stable conditions are *usually* associated with low ceilings, poor visibility, steady precipitation, rime icing, calm or steady wind, and no convective turbulence but possible mountain wave turbulence in mountainous areas.

The ability to recognize cloud types on satellite imagery can give a briefer a good idea of the stability of the atmosphere in a certain area and the type of hazards to look for.

In unstable air, uneven heating causes convective currents. Pockets or parcels of warm air rise and create updrafts. The rising air cools and clouds are formed in the updrafts. Between the updrafts, the air is sinking so no clouds exist. (See Fig. 1.)

Figure 1.

GOES Satellite Photo Recognition

PHOTO NUMBER TEN

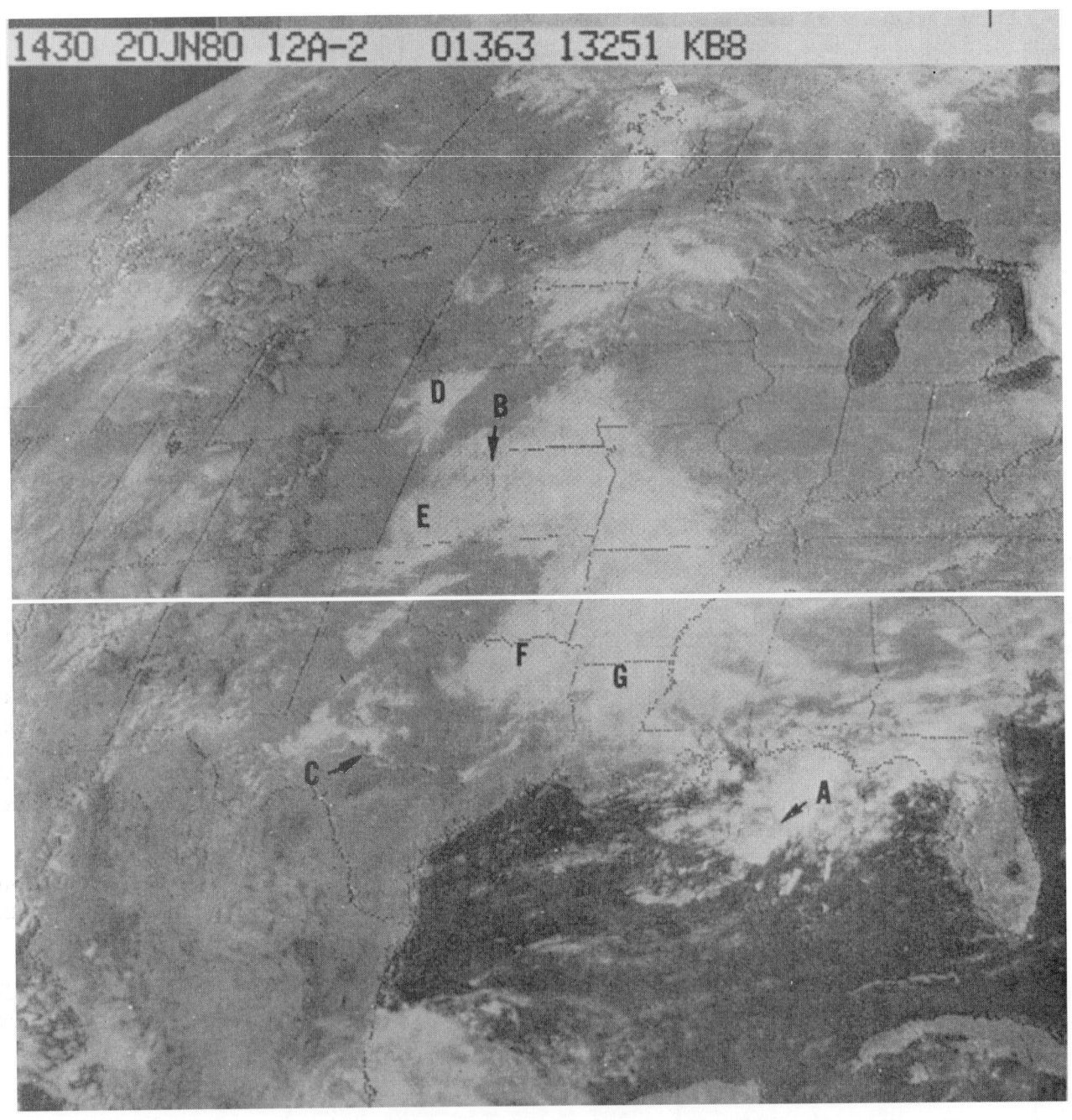

1-A-36

Often in unstable air, the tops of some cumulus clouds are higher than others. Early in the morning and late in the evening when the sun angle is low, the higher tops cast shadows on the lower ones. (See Fig. 2.)

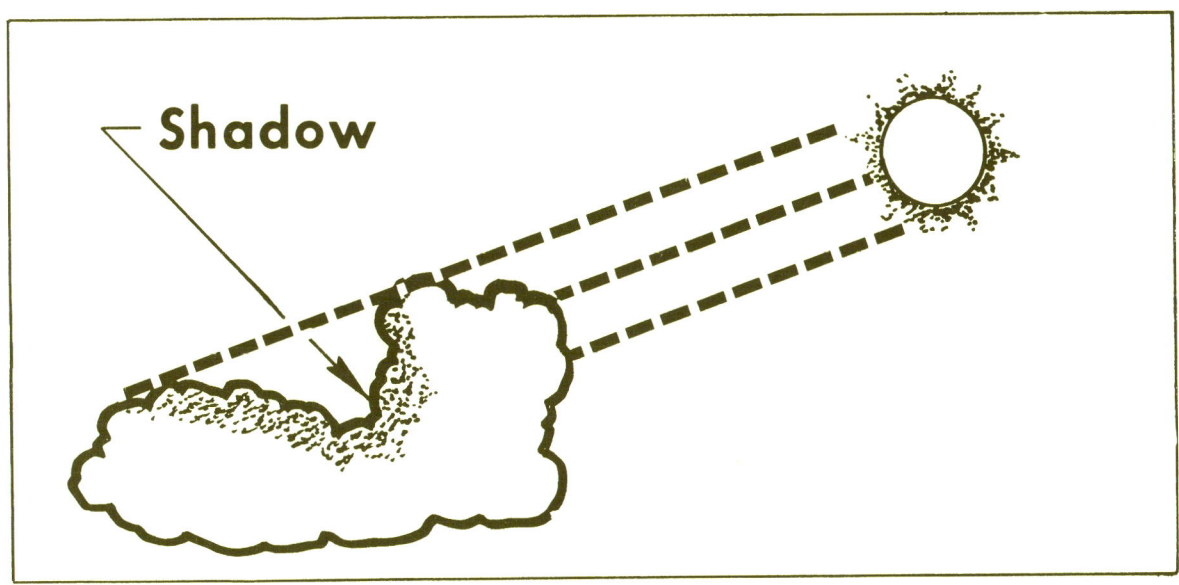

FIGURE 2.

These shadows can be seen on visible imagery and the resulting pattern is called texture.

Examples of texture can be seen at A and B, photo 10. Shadows can also be cast on the earth's surface, producing a similar effect (C).

At midday when the sun is directly overhead, no shadows are cast, and there will be no texture. Even when there are no shadows, cumuliform clouds may still be recognizable by their lumpy appearance, for example, the clouds around A. Sometimes, however, cumuliform clouds may look very much like stratiform clouds on a visible photo. Refer to photos 10 and 11 and note that on photo 10 the cirrus anvils of the thunderstorms at F and G look very much like the areas of fog and low stratus at D and E. In cases like this, you need an infrared photo or other aids such as surface observations and PIREPS to accurately determine cloud types.

Stratiform clouds normally do not have texture because of their flat tops; however, when there is more than one layer of clouds, the higher layer may cast shadows on the lower layer and produce the same effect. (See B, photos 10 and 11.)

Shadows do not appear on IR imagery so there is no texture, but sometimes there will be contours on the MB curve IR photo where the shadows appear on the visible photo (see A and B, photo 11). The contours show the edges of the higher clouds very clearly compared to the shade of the lower clouds.

GOES Satellite Photo Recognition

PHOTO NUMBER ELEVEN

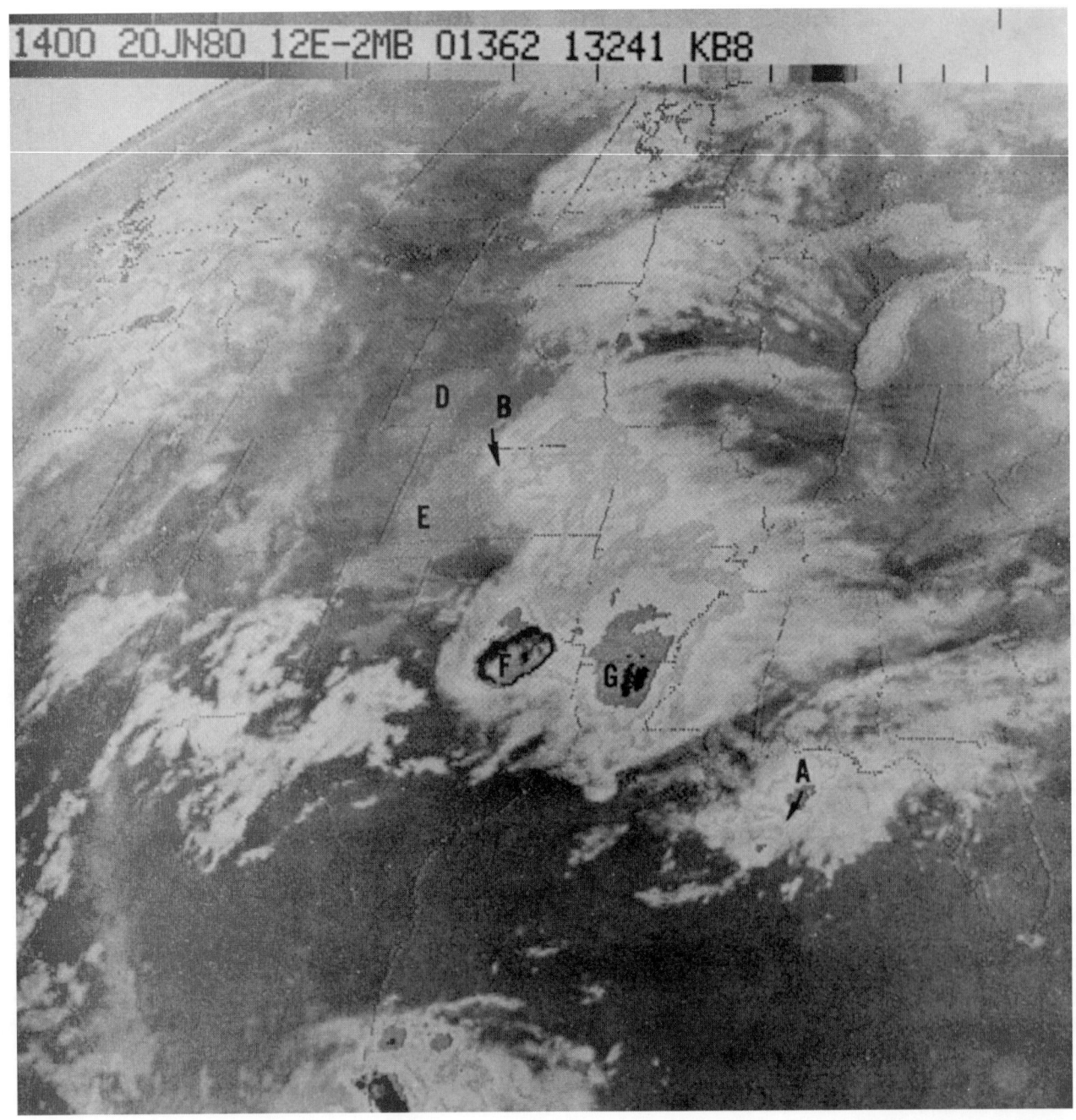

1-A-38

CRITERION TEST

CHAPTER II

1. Most clouds are formed in air that is cooled by

    a. contact with a cold surface
    b. mixing with colder air
    c. being lifted

2. How many basic cloud types are there and what are they called?

    a. (2) cumuliform and stratiform
    b. (3) high, middle, and low
    c. (2) thick and thin

3. Stratiform clouds are characteristically

    a. rounded and lumpy
    b. flat and sheetlike
    c. thin and wispy

4. Stratiform clouds are formed in

    a. descending air
    b. unstable air
    c. stable air

5. Cumulus clouds are formed in

    a. descending air
    b. unstable air
    c. stable air

6. Unstable conditions are usually associated with

    a. steady wind
    b. low ceilings and poor visibility
    c. convective turbulence

7. Stable conditions are usually associated with

    a. steady precipitation
    b. showery precipitation
    c. thunderstorms

8. Texture on satellite imagery is the result of

    a. warm areas that show up darker
    b. shadows on clouds from higher tops or layers
    c. openings between clouds that show darker terrain below

9. Texture can be seen on

    a. visible, unenhanced infrared, and enhanced infrared imagery
    b. visible and unenhanced infrared imagery only
    c. visible imagery only

10. Texture is best seen at

    a. sunrise
    b. midday
    c. midnight

GOES Satellite Photo Recognition

COMPREHENSIVE EXAM

NOTE: REFER TO PHOTOS 12 AND 13 TO ANSWER QUESTIONS 1 THROUGH 8. BE SURE TO USE ***BOTH*** PHOTOS FOR EACH QUESTION.

1. On photo 12, there are dark spots at A and B. Why are these spots darker than the surrounding area?

    a. They are warmer and, therefore, show up darker.
    b. Water surfaces reflect less sunlight than terrain surfaces.
    c. Heavily wooded areas reflect less sunlight than open rangeland.

2. On photo 12 there is a dark spot at C. Why is this spot darker than the surrounding area?

    a. It is warmer and, therefore, shows up darker.
    b. Water surfaces reflect less sunlight than terrain surfaces.
    c. Heavily wooded areas reflect less sunlight than open rangeland.

3. On photo 13, southeastern Oklahoma and most of the other southern states are a lighter shade than northwestern Oklahoma, Kansas, and Nebraska. What is the reason for the difference in shading?

    a. The lighter area is mostly covered by thin clouds.
    b. The darker area is heavily wooded and reflects less sunlight.
    c. Both areas are mostly clear; the shading represents surface temperatures.

4. The clouds at D are most likely

    a. mid-level clouds
    b. high, thin clouds
    c. low, thin clouds

5. The clouds at E are most likely

    a. mid-level clouds
    b. low, thin clouds
    c. high, thick clouds

6. On photo 13, A and B show up as lighter-shaded spots. This is because they are

    a. areas of tree-covered higher, cooler terrain.
    b. large bodies of cool water
    c. covered by low clouds

7. What type of clouds are found around E?

    a. Stratiform
    b. Cumuliform
    c. Cannot be determined

GOES Satellite Photo Recognition

8. What aviation hazard would pilots and briefers more likely be concerned with at an airport in the vicinity of E?

   a. Widespread low ceilings and steady rain.
   b. surface visibility restricted due to fog.
   c. low-level wind shear from thunderstorm gust fronts.

NOTE: REFER TO PHOTOS 14 and 15 TO ANSWER QUESTIONS 9 THROUGH 15.

9. What type of cloud is located at A?

   a. Stratiform
   b. Cumuliform
   c. Cannot be determined

10. To what family does this cloud belong?

    a. High
    b. Middle
    c. Low

11. What aviation hazard would pilots and briefers more likely be concerned with in this area?

    a. Convective turbulence
    b. Low ceiling and restricted surface visibility
    c. Gusty surface wind

12. On photo 15, the area around B is a definite shade lighter than around C and D. Why?

    a. The area around B is covered by low clouds.
    b. The sea surface at B reflects more sunlight so it appears lighter.
    c. The sea surface temperature at B is cooler than at C and D.

13. The dark spots at E are

    a. islands protruding through low clouds.
    b. the sea surface seen through holes in the clouds.
    c. shadows on low clouds from higher clouds.

14. On photo 14, the dark line at F is a

    a. large river.
    b. heavily wooded mountain ridge.
    c. shadow from the high cloud just to the west.

15. Select the most likely statement concerning the clouds at G and H. The clouds at G are

    a. low; at H, mid-level
    b. high; at H, low
    c. mid-level; at H, mid-level

1-A-41

GOES Satellite Photo Recognition

PHOTO NUMBER TWELVE

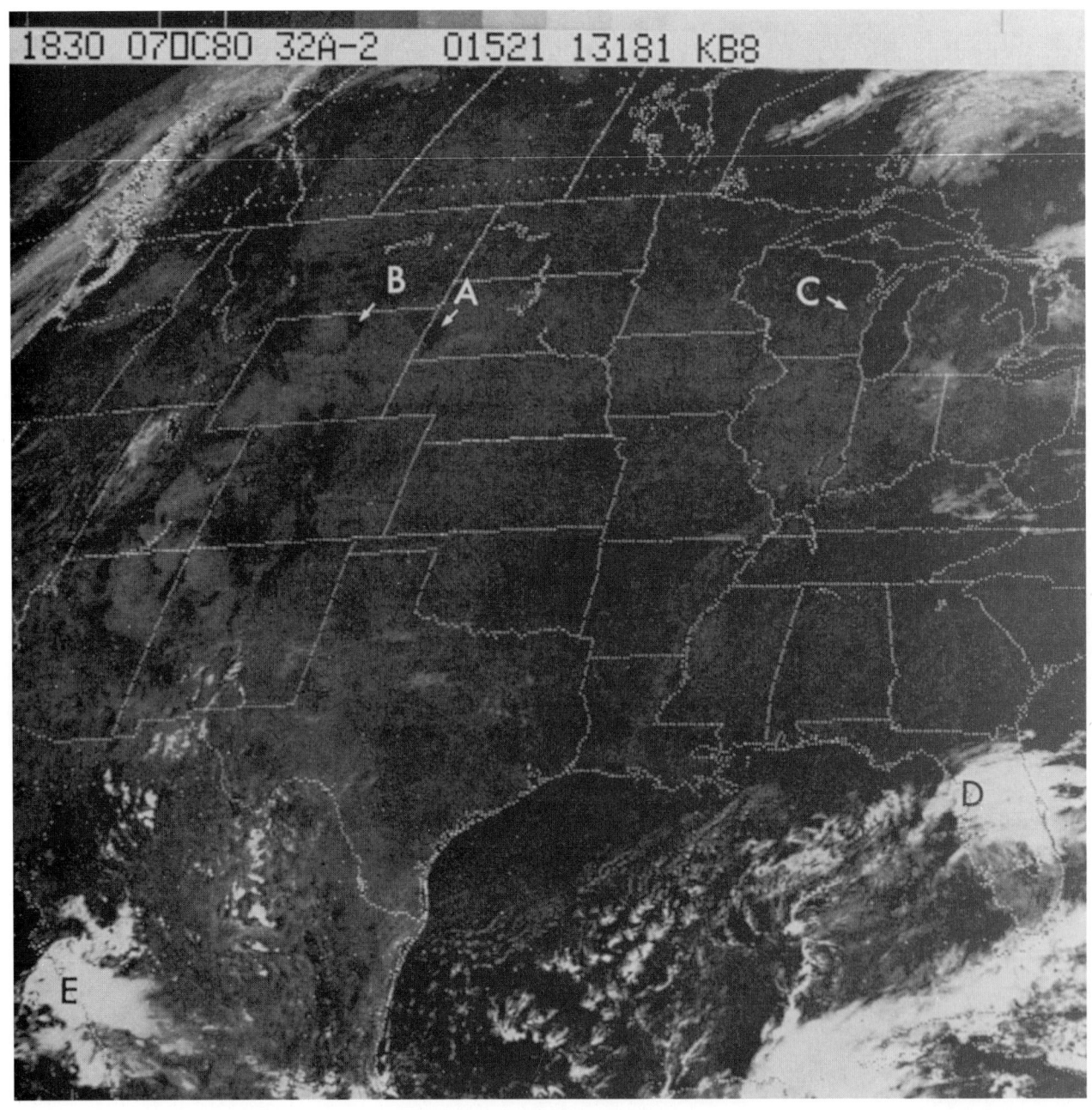

1-A-42

GOES Satellite Photo Recognition
PHOTO NUMBER THIRTEEN

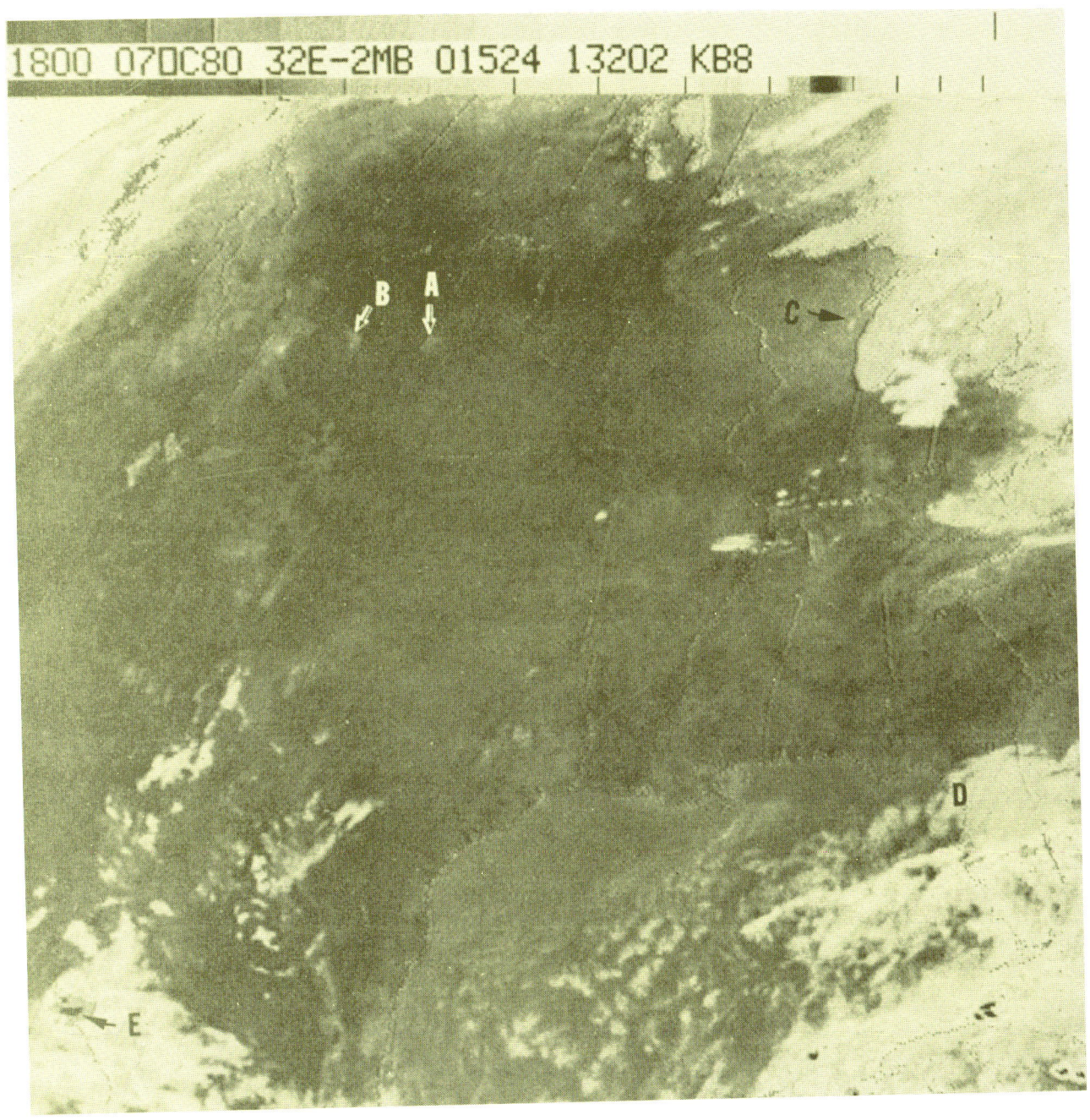

1-A-43

GOES Satellite Photo Recognition
PHOTO NUMBER FOURTEEN

1-A-44

GOES Satellite Photo Recognition
PHOTO NUMBER FIFTEEN

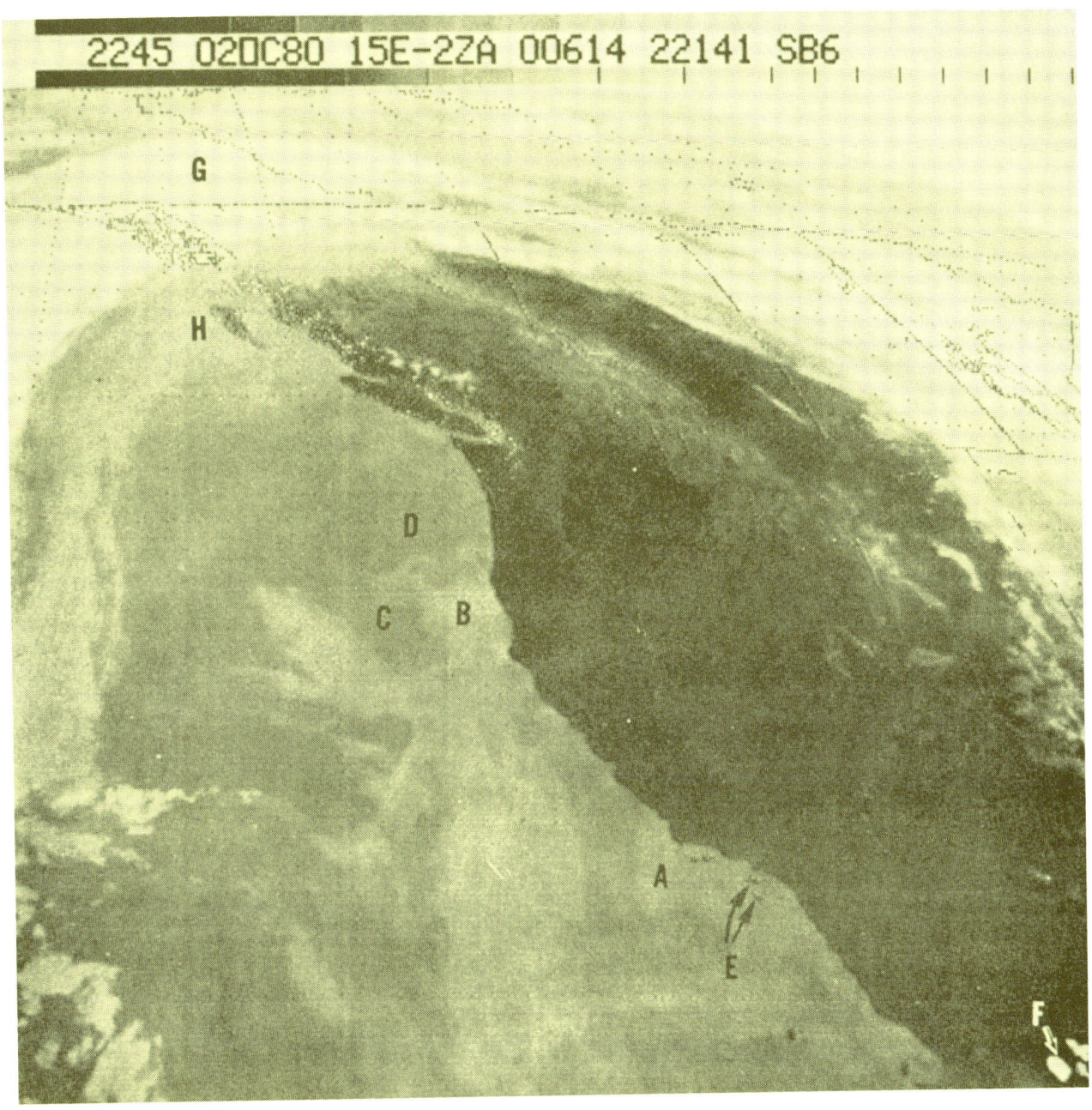

GOES Satellite Photo Recognition

CRITERION TEST 1

Answer	Page Reference
1. a	7
2. c	7
3. c	7
4. b	17
5. b	16-17
6. c	9-13
7. b	17-19
8. c	7
9. a	18-21
10. b	7
11. c	26-29
12. a	7
13. c	18 and 26
14. a	15
15. c	15 and 29
16. b	15

CRITERION TEST 2

Answer	Page Reference
1. c	35
2. a	35
3. b	35
4. c	35
5. b	35
6. c	35
7. a	35
8. b	37
9. c	37
10. a	37

COMPREHENSIVE EXAM

Answer	
1. c	Heavily forested Black Hills of SD and Bighorn Mountains of WY.
2. b	Lake Winnebago, WI.
3. c	Visible photo shows no clouds, IR shows temperature difference.
4. a	Medium gray shade. High clouds, even thin, would be whiter, low thin clouds would be much darker.
5. c	MB contouring indicates very cold (high) tops.
6. a	Higher elevation plus tree cover keeps mountainous areas cooler.
7. b	MB photo shows overshooting top of a thunderstorm.
8. c	Low-level wind shear results from downrush of air in a thunderstorm.
9. a	Stratiform clouds appear flat and sheetlike, no texture.
10. c	Dark shading on IR indicates quite warm, therefore low.
11. b	Stable conditions often cause low ceilings and poor visibility.
12. c	Visible photo shows no clouds, IR shows significant temperature difference.
13. a	IR photo shows the spots much darker (warmer) than the open sea surface.
14. c	IR photo confirms high cloud to the west, afternoon sun casts a shadow to the east.
15. b	IR photo shows white shade at G, dark shade at H.

PICTURE OF THE MONTH

Reliability of Enhanced Infrared (EIR) Geostationary Satellite Data at High Latitudes

FRANCES C. PARMENTER-HOLT[1]

Environmental Products Branch, NOAA, National Earth Satellite Service, Washington, DC 20233

(Manuscript received 26 February 1982, in final form 24 June 1982)

An increasing need for quantitative values from satellite data has led to the development of numerous infrared data enhancement curves (Corbell *et al.*, 1976). Most of these curves have been designed for use on the geostationary (GOES) satellite data and represent temperatures typical of the tropical to mid-latitude regions that lie in the main field of view of these satellites.

The utility of continuous meteorological satellite data for high-latitude areas, such as Alaska, soon became apparent. Early in 1979, the Anchorage Satellite Field Services Station of NOAA's National Earth Satellite Service (NESS) began receiving a full 24 hours per day program of GOES West sectors. Low sun angles for nearly a third of the year require extensive use of infrared (IR) data in Alaska. Thus, new IR enhancement curves that addressed the temperatures observed in this region were developed. After a year of operational use, it was found that realistic quantitative temperature values could be obtained from the geostationary enhanced IR data.

All of Alaska lies north of 50°N, far from the equatorial subpoint of the GOES West satellite and well into the "near-horizon" viewing area of the spacecraft. Data interpretation problems, such as cloud displacement, near the horizon of the GOES data have been addressed by Weiss (1978). Water vapor and other atmospheric constituents absorb outgoing radiation and can result in "cooler" than expected satellite observed temperatures. This effect, discussed by Anderson *et al.* (1974) in regard to polar orbiting satellite data is accentuated in near-horizon areas where the sensor is looking obliquely through a long atmospheric path. This study examines GOES

---

[1] Formerly Manager, Satellite Field Services Station, Anchorage, AK.

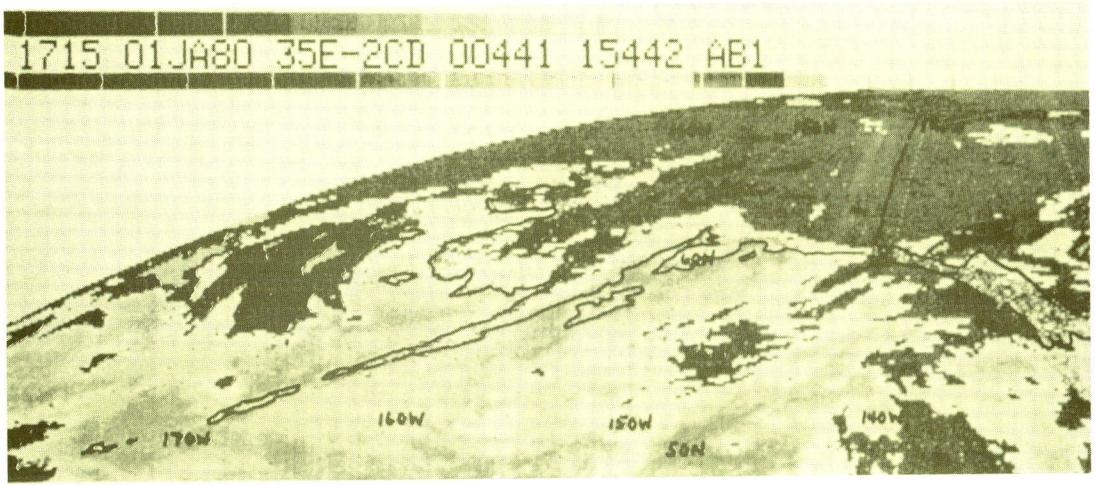

FIG. 1. GOES W, EIR, CD curve 1715 GMT, 1 January 1980.

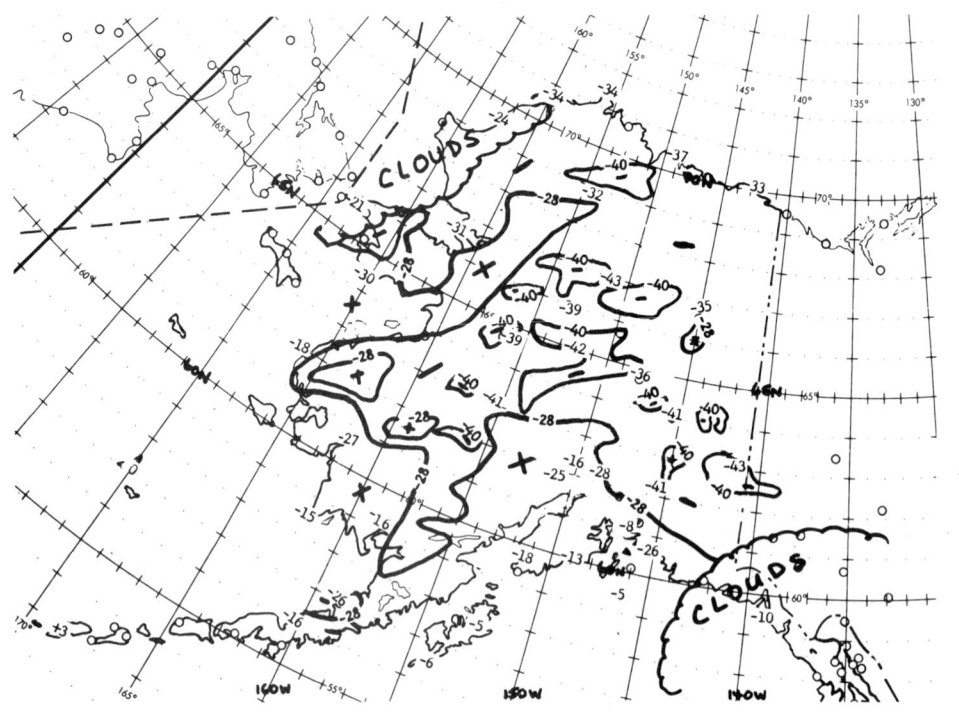

FIG. 2. Satellite-derived isotherms (solid lines) from Fig. 1 and observed surface temperatures.

satellite derived and conventionally observed temperatures for Alaska for two winter cases taken from the operational files.

A large area of high pressure brought clear skies and light winds to much of central Alaska on 1 January 1980. The GOES West enhanced IR (EIR) view of the state appears in Fig. 1. These data are displayed with the CD curve; temperature and gray shade scales for this curve appear below the date/time documentation line. The gray shading in this CD enhancement curve progresses from black, representing the warmest temperatures at the left end, toward light gray until

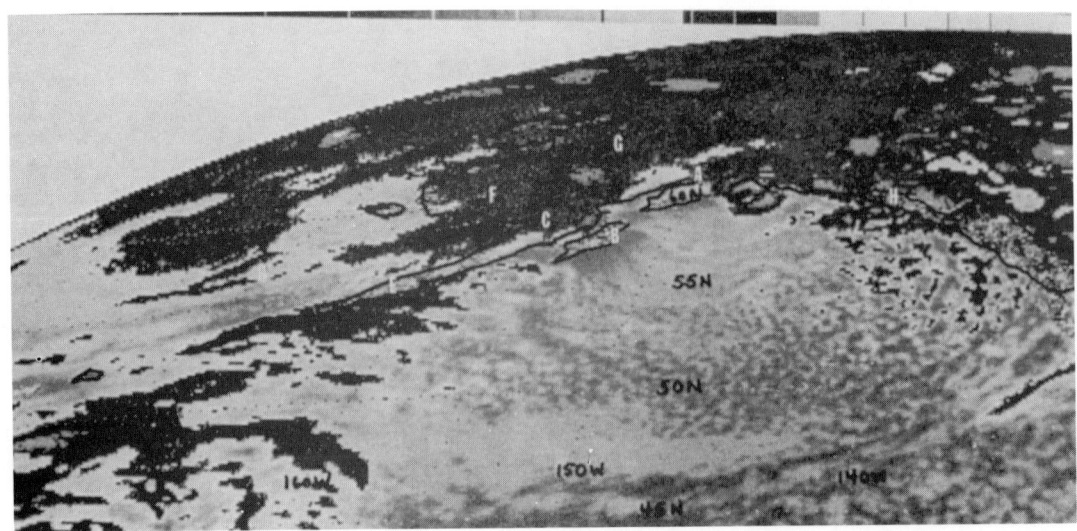

FIG. 3. GOES W, EIR, CD curve 1215 GMT, 16 January 1980, with radiosonde stations marked.

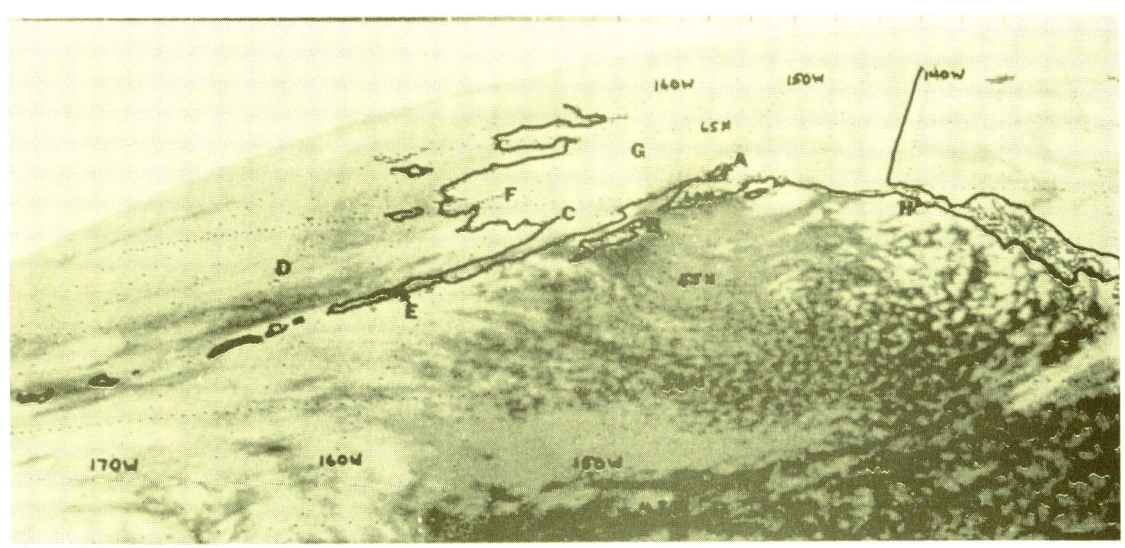

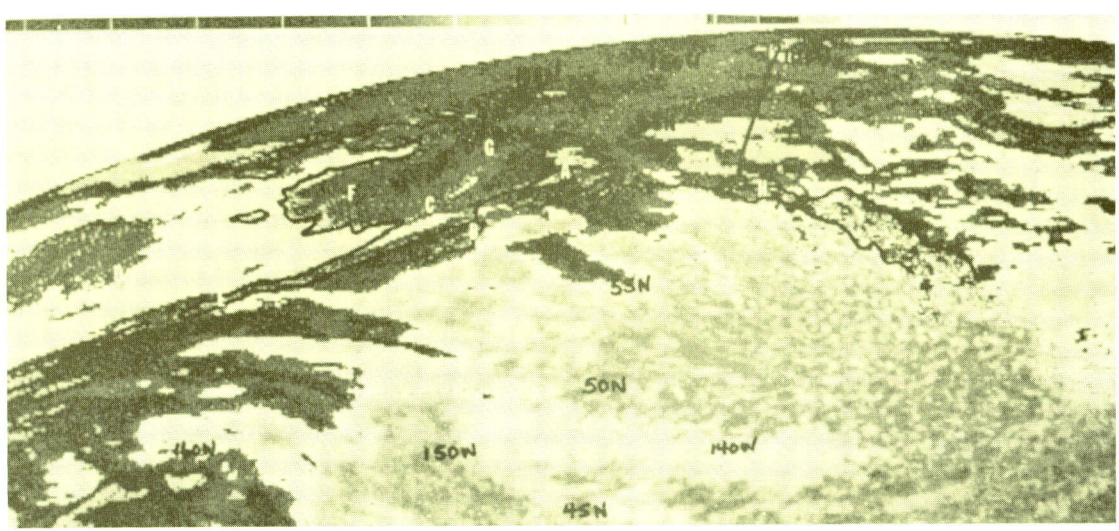

FIG. 5. As in Fig. 3, but for 0115 GMT, 17 January 1980.

temperatures reach $-28°C$.[2] Then alternating darker tones are used: dark gray depicts -28 to $-40°C$; black -40 to $-50°C$; medium gray -50 to $-60°C$; white, colder than $-60°C$. The -28 and $-40°C$ isotherms, analyzed directly from this EIR image (Fig. 1), appear as solid lines in Fig. 2. (Plus and minus signs indicate whether the area enclosed is warmer or colder, respectively.) Although the CD curve temperature intervals are large, the satellite-derived isotherms are in good agreement with the observed 1800 GMT surface temperatures also appearing in Fig. 2. This is true even at $70°N$, along the north coast of the state.

TABLE 1. Eight radiosonde locations.

A	Anchorage (273)
B	Kodiak (350)
C	King Salmon (326)
D	Saint Paul Island (308)
E	Cold Bay (316)
F	Bethel (219)
G	McGrath (231)
H	Yakutat (361)

---

[2] Each vertical tick mark represents $10°C$. The $-28°C$ gray shade begins just to the left of the AB1 label in the documentation in Fig. 1.

FIG. 6. As in Fig. 3, but for 1215 GMT, 17 January 1980.

A series of images, taken between 1215 GMT 16 January and 1215 GMT 17 January 1980 are shown in Figs. 3–6. During this time, a weak low was moving into southeast Alaska and cloudiness was increasing in the south central and southwestern portions of the state. Data in Figs. 3, 5 and 6 are displayed with the same CD curve used in Fig. 1. A regular IR display appears in Fig. 4.

Cloud-top temperatures at eight radiosonde locations (Table 1) were derived from the satellite imagery. These temperatures appear in Table 2; their locations are noted on Figs. 3–6. [No viewing angle corrections (Weiss, 1978) were used to relocate the cloud areas.] Table 2 also summarizes the cloud-top temperatures determined from the radiosonde observations. Dewpoint data, necessary to define cloud tops, often terminated before the top of a cloud layer could be determined. For these cases (with asterisks), the coldest temperature at which the dewpoint spread suggested a cloud layer was used. Where a complete data set was available, the satellite EIR temperatures compared well with conventional data.

GOES satellite data provide good areal coverage of central and southern Alaska. EIR data appear to reliably depict both surface and cloudtop temperatures throughout the year. By using radiosonde and satellite observations, the analyst can assign heights to the observed cloud cover. Changes in cloud cover and cloud height, particularly thunderstorm tops, can easily be monitored from the half-hour interval GOES imagery and are regularly included in the products of the Anchorage SFSS.

One of the users of these products is the aviation community. Aviation is the primary means of transportation in Alaska; vital medical, postal and supply services to many locations can only be provided by

TABLE 2. Cloud-top temperatures (°C) derived from satellite imagery at eight radiosonde stations.

Station	1200 GMT, 16 January 1980		0000 GMT, 17 January 1980		1200 GMT, 17 January 1980	
	Radiosonde	Satellite	Radiosonde	Satellite	Radiosonde	Satellite
A	−31*	−28 to −40	−30	−28 to −40	−40*	−50 to −60
B	−10.5**	−10 to −15	−9, −37	−28 to −40	−40*	−50 to −60
C	−28.9	−28 to −40	−40	−28 to −40	Missing	−50 to −60
D	−38	−28 to −40	Missing dewpoint	−20 to −28	−28** 2° Inversion	−20 to −28
E	−35	−28 to −40	−39	−28 to −40	−38*	−40 to −60
F	−40	−28 to −40	Missing	−28 to −40	−37.9**	−40 to −50
G	−40*	−40 to −50	−40*	−40 to −50	−30*	−40 to −60
H	−8, −45*	−40 to −50	Missing	−10 to −28 −40 to −50	−41*	−40 to −50

* Temperature of last saturated condition observed before dewpoint reading was lost.
** Clear, temperature at surface.

air. Most observations are made during the day, and little weather information is available through the night until pilots are airborne. Thus, the quantitative information on cloud cover, cloud height and system movement available from the GOES satellite data is particularly useful to the aviation forecaster and the pilot.

REFERENCES

Anderson, Ralph K., *et al.*, 1974: Applications of meteorological satellite data in analysis and forecasting. ESSA Tech. Rep. NESC 51, National Earth Satellite Service, Washington, DC 20233, pp. 6c 1–18.

Corbell, Ralph P., J. Cornelius Callahan and William J. Kotsch, 1976: *The GOES/SMS User's Guide*. National Oceanic and Atmospheric Administration/National Earth Satellite Service and National Aeronautics and Space Administration, Washington, DC 20233, 80 pp.

Weiss, Carl E., 1978: Cloud-location corrections near the horizon of an SMS image. Satellite Applications Interpretation Note 78/8, National Earth Satellite Service, Washington, DC 20233, 8 pp.

TECHNICAL ATTACHMENT 77-G4

DISPLACEMENT ERROR OF SATELLITE CLOUD TOPS

By using enhanced IR satellite data and other types of information, we can locate the highest tops of thunderstorms in satellite pictures. However, to use this information to the fullest, especially for hydrologic purposes, we need to know quite precisely the location of the top on a map of the area. Directly below the satellite this is simple, since the top, as seen via satellite, is directly above its true position on the ground. This, of course, is only true at the equator at the longitude of the satellite. As the cloud gets farther from the equator and/or farther from the longitude of the satellite, our view is more and more of the side of a tall cloud, and an error in the location of the top is introduced. This error depends also on the height of the top.

If the earth were flat, the error would be brought about as in Figure 1, for a cloud away from the equator and at the longitude of the satellite ($75°$ for GOES-1) where the top would appear to be at point $\underline{a}$ instead of its proper location at point $\underline{b}$. However, because of the great height of the satellite, the amount of the error for the distance of Central Region locations from the sub-satellite point ($0°$ lat. and $75°$ lon. for GOES-1) would be small, because the angle of concern (α_f) would be less than $8°$ and magnitude of the error would be under one nautical mile for a 40,000 foot cloud top.

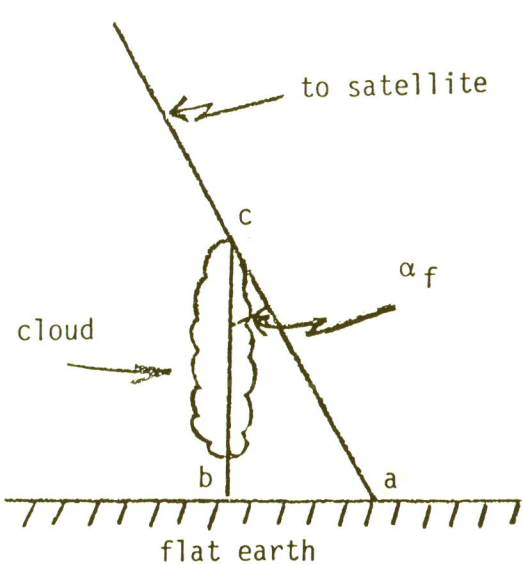

Figure 1

For the round earth, the angle of concern increases by the amount of the latitude itself, as seen in Figure 2, i.e. ($\alpha_r = \alpha_f +$ the latitude), so the error increases with latitude (distance from the satellite subpoint)[1].

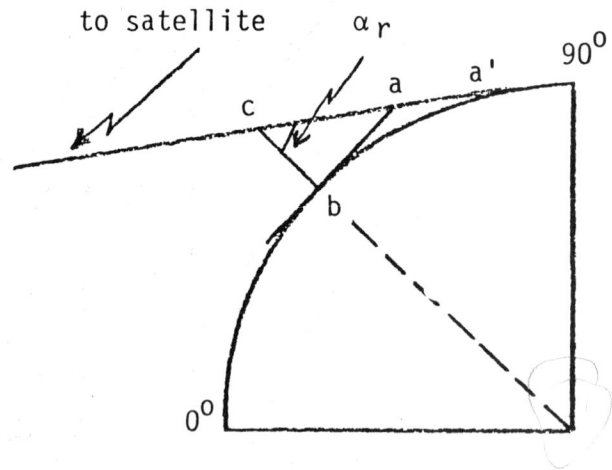

Figure 2

It turns out that the lines of equal displacement error are concentric circles around the sub-satellite point on a full disc picture and thus are portions of circles on any portion of the full disc, as our pictures are. The lines of equal error are given on the attached map for a 40,000 foot top.[2] Note that they are given as nautical miles.[3] Adjustment for tops of other heights is easy, as the error is essentially a linear function of height, and would be 50% greater for a 60,000 ft. top and 50% less for a 20,000 ft. top. To correct a top location, one should move it the amount of the error in the direction of the arrows shown, i.e., perpendicular to the lines and toward the satellite.

For SMS-2 pictures, the magnitude of the error for the longitude of Denver (105°W) is the same since that is the mid-longitude between the two satellites. The error east (west) of there would be larger (smaller) than at 105°W. The lines of equal SMS-2 error would be nearly perpendicular to those in Wyoming and Colorado, so the adjustment would be by a movement toward about the southwest in those states.

---

[1] Also the point a on the plane tangent at point b becomes a' on the curved earth, so a'b is greater than ab. However, for the cloud heights possible with convective systems, and for the farthest distance away from the sub-satellite point in the Central Region, the difference between a'b and ab is negligible and is neglected here.

[2] This is a corrected version of the map distributed at the satellite workshop in Kansas City in February.

[3] To change nautical miles to statute miles, multiply by 1.15, or add 15%.

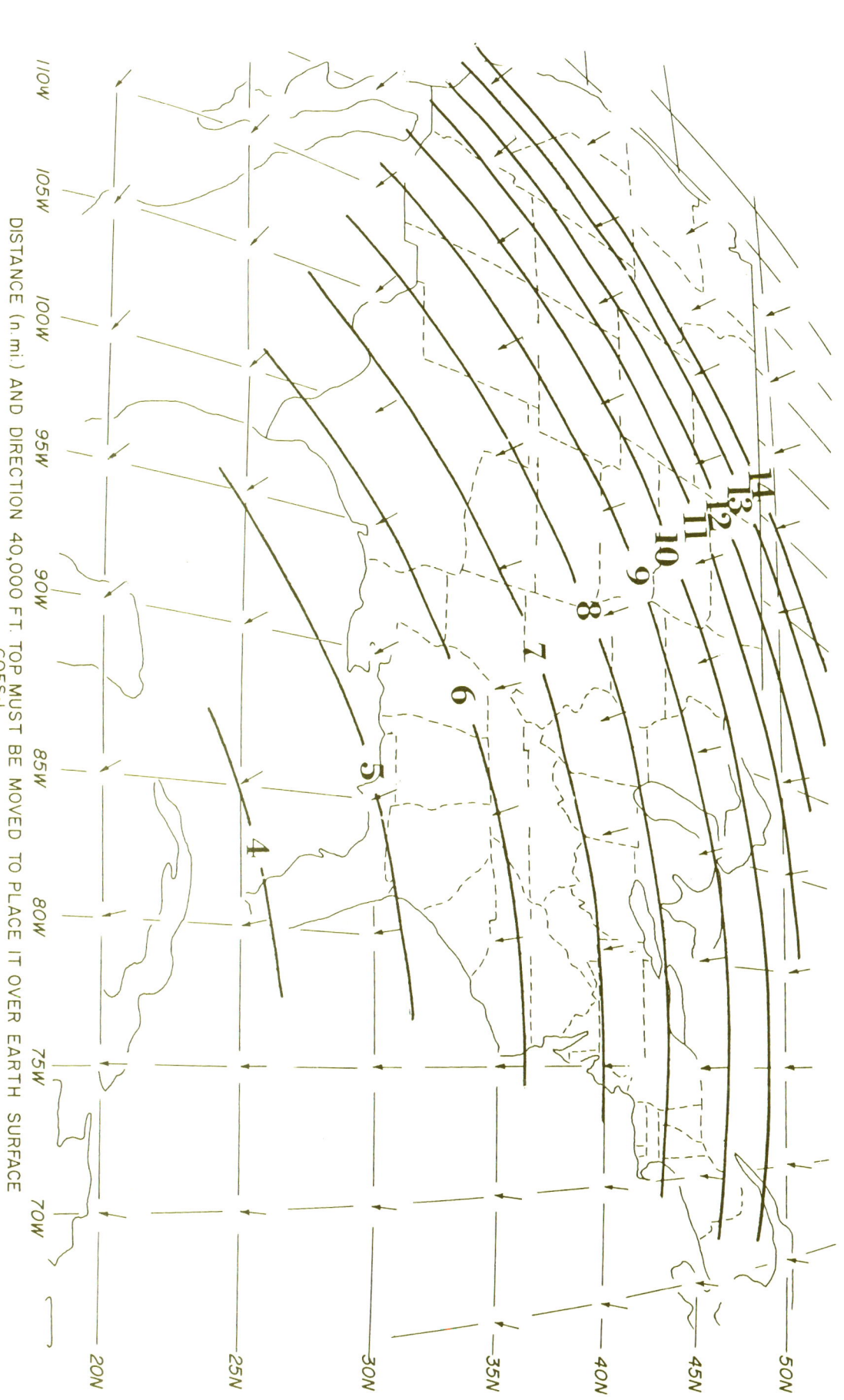

U.S. DEPARTMENT OF COMMERCE

National Weather Service/National Environmental Satellite Service
SATELLITE APPLICATIONS INFORMATION NOTE 78/8

CLOUD-LOCATION CORRECTIONS NEAR THE HORIZON
OF AN SMS IMAGE

Carl E. Weiss
Applications Division, NESS, Washington, D.C.

Cloud interpretation near the horizon of an SMS image is, in part, complicated by the fact that the earth's surface curves rapidly away from the satellite line-of-sight. For a cloud located precisely over the satellite subpoint, the apparent and actual cloud positions coincide, but the difference between the apparent and actual positions increases as the cloud's distance from the subpoint increases.

Figure 1 illustrates this problem. In order to accurately locate a cloud feature on an SMS image, a correction for the effect of the earth's curvature must be made in the direction of the satellite subpoint. Near the horizon, this correction will be large. This SAIN gives cloud location corrections for areas near the edge of the earth's disk.

Several authors have addressed this problem and have calculated cloud location corrections primarily for areas within 50° great circle arc of the satellite subpoint. Pike (1974) developed latitude/longitude component corrections for cloud tops of 40,000 feet over the Atlantic and Caribbean. National Weather Service Central Region Headquarters (1977) calculated location corrections over the U.S. and displayed them as concentric circles of equal correction centered on the satellite subpoint. Corrections were given for cloud tops of 20,000 feet, 40,000 feet, and 60,000 feet.

The cloud-location corrections developed in this note apply to areas of 50° great circle arc and farther from the satellite subpoint. Also, the corrections can be applied to cloud tops of any height. The dimensions used for determining the cloud corrections are shown in Figure 2. Here ϕs is the great circle arc from the satellite subpoint to a cloud's apparent location on an SMS image, h is the cloud height above the earth's surface, $\bar{R}$ is the mean earth radius, $\bar{R}_e$ is the mean equatorial earth radius, $\bar{A}$ is the mean SMS altitude, and d is the location correction[1]. A sample calculation for a cloud top at 10 km and appearing at 60° of arc is given in the appendix.

Cloud-location corrections were computed at 5° great circle arc intervals, starting at 50° of arc and ending at 70° of arc. At each arc interval, corrections were determined for cloud heights of 3 km, 10 km, and 15 km. These corrections are shown in Table 1. We see that for a cloud top at 5 km and appearing at 55°, the cloud position must be corrected by 9.67 km toward the satellite subpoint.

---

[1]. This correction represents a distance measured along the curved surface of the earth.

Now, by taking the ratios of the location corrections to the cloud heights (d/h) at a given ϕs, we find that the results are, for practical purposes, the same. The mean d/h ratio values or coefficients at each ϕs can be used to find corrections for clouds of any height. Multiplying the cloud height by the appropriate $\overline{d/h}$ value from Table 2 gives the distance by which the cloud's position must be corrected. Because the ratio values are dimensionless, they can be used with cloud heights given in any units. Remember, these location corrections are always applied radially toward the satellite subpoint.

Mean d/h values were plotted against the appropriate great circle arc (Figure 3). Then ϕs values were interpolated for each 0.5 interval of $\overline{d/h}$ coefficient. Each $\overline{d/h}$ coefficient was drawn on an SMS grid as a concentric circle corresponding to the appropriate great-circle arc (Figure 4). Radial lines converging at the subpoint indicate the direction of the correction.

As an example of using Figure 4, consider a cloud top at 10 km appearing over Anchorage (ANC). This position must be corrected by 30 km to the SSE to find the actual cloud location. Corrections for clouds appearing between circles can be estimated by interpolating the value of the correction coefficient between circles.

The correction coefficients provide a quick and easy method for determining cloud location corrections for cloud tops of any height. Nomograms such as Figure 4 can be drawn for various locations viewed by either SMS to give the user a handy means to accurately locate cloud features.

REFERENCES

NWS Central Regional Technical Attachment No. 77-G4, April 1977, Displacement Error of Satellite Cloud Tops.

Pike, A. C., 1974: A Short Routine for the Adjustment of Geostationary Satellite Cloud Positions Viewed Obliquely. Unpublished Paper, NWSS, Coral Gables, Fla., 8 pp.

APPENDIX

A sample cloud displacement calculation for a cloud top at 10 km and appearing at 60° of great circle arc from the SMS subpoint is shown below. Referring to Figure 2, the following information is known:

$\overline{R}$ = 6371 km

$\overline{R}_e$ = 6378 km

ϕs = 60°

$\overline{A}$ = 35,793 km

h = 10 km

For triangle $\gamma\beta\phi s$, we can write

(1) $\tan 1/2(\beta - \gamma) = \dfrac{\overline{R} - (\overline{R}_e + \overline{A})}{\overline{R} + (\overline{R}_e + \overline{A})} \tan 1/2(\beta + \gamma).$

However, substituting the given values into (1), we have

(2) $\dfrac{\overline{R} - (\overline{R}_e + \overline{A})}{\overline{R} + (\overline{R}_e + \overline{A})} = -\dfrac{35800 \text{ km}}{48542 \text{ km}} = -0.737506.$

Remembering that the sum of the interior angles of any triangle equals 180°, we have for triangle $\gamma\beta\phi s$

(3) $\beta + \gamma + \phi s = 180°.$

Solving for $\beta + \gamma$

(4) $\beta + \gamma = 180° - \phi s.$

Substituting (3) and (4) into (1), we can write

(5) $\tan 1/2(\beta - \gamma) = -0.737506 \tan 1/2(180° - \phi s).$

We are given that $\phi s = 60°$, so

$\tan 1/2(\beta - \gamma) = -0.737506 \tan 1/2(180° - 60°)$

or,

$\tan 1/2(\beta - \gamma) = -0.737506 \tan 1/2(120°)$
$= -0.737506 \tan 60°$
$= (-0.737506)(1.732051)$

(6) $\tan 1/2(\beta - \gamma) = -1.277397.$

Evaluating (6) gives us

$$\frac{1}{2}(\beta - \gamma) = \text{Arctan}(-1.277397)$$
$$\frac{1}{2}(\beta - \gamma) = -51.944669°,$$

and

(7) $\quad \beta - \gamma = -103.889338°.$

Now, from (4) we have

(8) $\quad \beta + \gamma = 120°,$

and adding (8) to (7) gives

$$\begin{aligned} \beta - \gamma &= -103.889338° \\ \beta + \gamma &= 120° \\ \hline 2\beta &= 16.110662°, \end{aligned}$$

(9) $\quad \beta = 8.055331°.$

Solving (4) for γ, we have

(10) $\quad \gamma = 180° - (\phi s + \beta).$

Substituting the given values into (10) gives

$$\begin{aligned} \gamma &= 180° - (60° + 8.055331°) \\ &= 180° - 68.055331° \end{aligned}$$

(11) $\quad \gamma = 111.944669°$

The law of sines gives us for triangle $\delta\gamma\eta$, Figure 2,

(12) $\quad \dfrac{h + \overline{R}}{\sin \gamma} = \dfrac{\overline{R}}{\sin \delta}$

Solving (12) for $\sin \delta$, we have

(13) $\quad \sin \delta = \dfrac{\overline{R} \sin \delta}{h + \overline{R}}$

Evaluating (13) gives

$$\sin \delta = \frac{(6371 \text{ km}) \sin(111.944669°)}{10 \text{ km} + 6371 \text{ km}}$$
$$= \frac{(6371 \text{ km})(0.927545)}{6381 \text{ km}}$$

$\sin \delta = 0.926022,$

and

(14) $\quad \delta = \text{Arcsin}(0.926022) = 67.833535°.$

For triangle $\delta\gamma\eta$, summing the interior angles we have

(15) $\delta + \gamma + \eta = 180°$.

Solving (15) for η and evaluating gives

$$\begin{aligned}\eta &= 180° - (\delta + \gamma) \\ &= 180° - (67.833535° + 111.944669°) \\ &= 180° - 179.778204°\end{aligned}$$

(16) $\eta = 0.221796°$.

Finally, the arc length d is given by

(17) $d = \dfrac{\pi \overline{R} \eta}{180°}$.

Evaluating (17), we have

(18) $d = \dfrac{\pi(6371 \text{ km})(0.221796°)}{180°} = 24.662590$ km.

Therefore, for a cloud top at 10 km and appearing at 60° of arc from the SMS subpoint, the displacement is 24.66 km toward the subpoint.

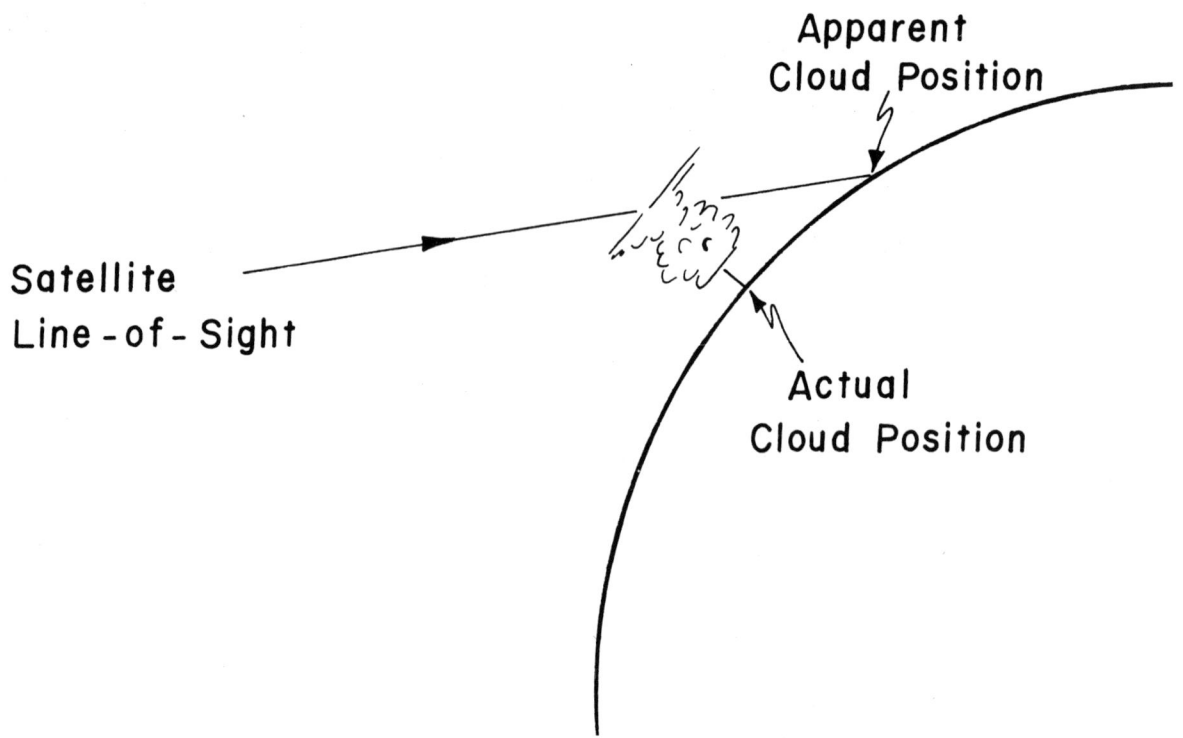

Figure 1. Diagram illustrating an actual vs. an apparent cloud location.

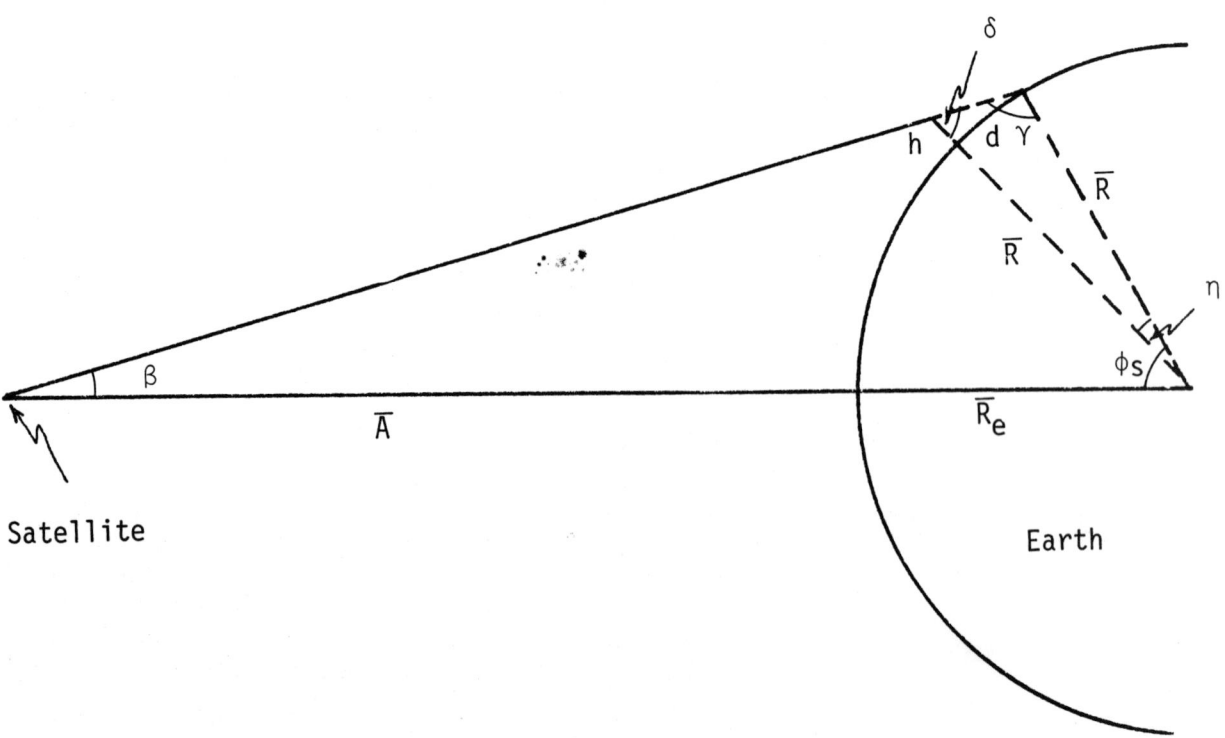

Figure 2. Dimensions used in calculating cloud location corrections.

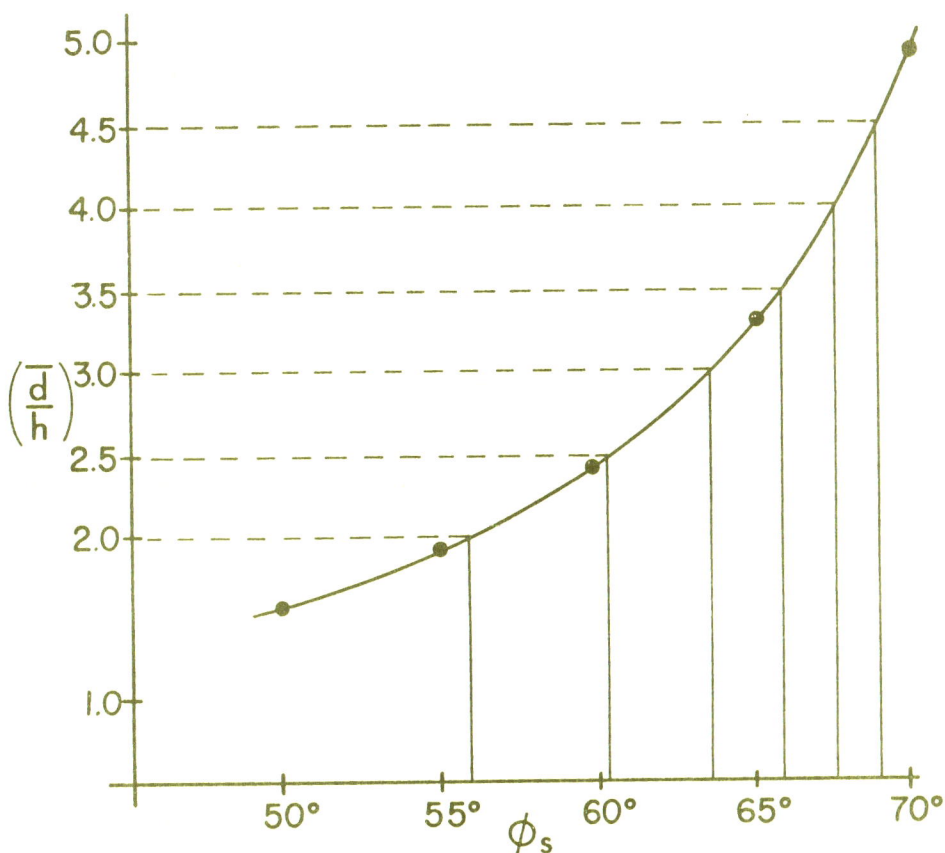

Figure 3. Mean correction coefficients, $(\overline{\frac{d}{h}})$, versus great-circle arcs, ϕ_s.

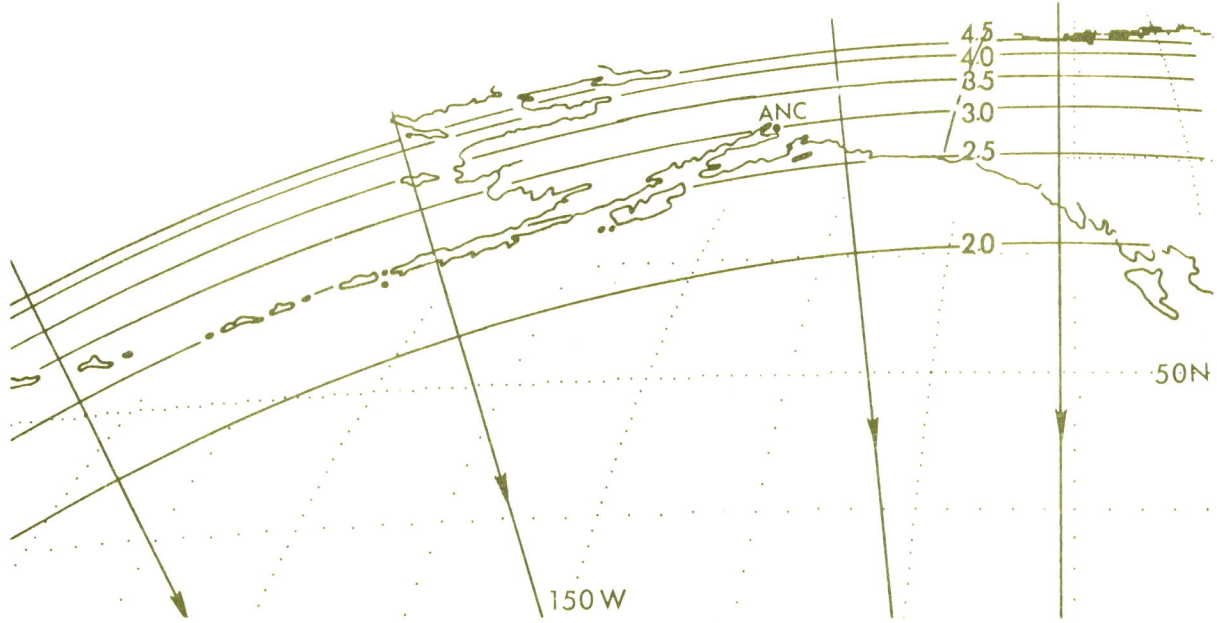

Figure 4. Mean correction coefficients, $(\overline{\frac{d}{h}})$, at 0.5 unit intervals plotted as concentric circles about the SMS-2 subpoint.

$\phiس$ h	50°	55°	60°	65°	70°
3km	4.67	5.81	7.43	9.98	14.67
5km	7.78	9.67	12.37	16.60	24.36
10km	15.53	19.30	24.66	33.04	48.24
15km	23.25	28.89	36.88	49.31	71.66

Table 1. Cloud-location corrections (d) in kilometers for selected cloud heights at given ϕs values.

ϕs h	50°	55°	60°	65°	70°
3km	1.56	1.94	2.48	3.33	4.89
5km	1.56	1.93	2.47	3.32	4.87
10km	1.55	1.93	2.47	3.30	4.82
15km	1.55	1.93	2.46	3.29	4.78
$(\overline{\frac{d}{h}})$	1.55	1.93	2.47	3.31	4.84

Table 2. Correction coefficients (d/h) of cloud-location for selected cloud heights and average coefficients at given ϕs values.

SUPPLEMENTARY INFORMATION FROM NESDIS FILES

I. <u>NOAA, or Polar Orbiting Weather Satellites.</u>

   A. Background.

The NOAA spacecraft is a five-sided structure 3.7 meters long, 1.8 meters wide, and weighing about 1410 kilograms. At one end of the spacecraft is a solar panel which is continuously motor-driven to face the sun during daylight portions of the orbit. The NOAA spacecraft images continuously, even when it is in eclipse.

The first satellite of this sort was launched February 3, 1966 from Cape Canaveral. The current configuration of these spacecraft began with the launch of TIROS-N in October, 1978. Two NOAA spacecraft normally image each point on the earth four times a day.

NOAA satellites are launched with a 98° inclination to the pole so that the orbit is sun-synchronous and precesses 360° in 365 days. The satellite is always in the same orientation with respect to the sun. One satellite descends (moves south) over every area on earth about 02:45 local time and ascends about 14:45 local time. The other satellite descends about 07:30 local time, and ascends about 19:30 local time. The orbital altitude is 833 kilometers, and each orbit is complete after 101.3 minutes.

   B. NOAA Sensor Systems.

      1. Primary NOAA Sensor.

The primary NOAA sensor is the Advanced Very High Resolution Radiometer (AVHRR). AVHRR detectors respond to radiation reflected or incident from the earth within five specific intervals (channels) of the electromagnetic spectrum to produce the image.

<u>CHANNEL 1</u> records visible radiation, with wavelengths of 0.58 to 0.68 microns. This channel is used for detection of cloud cover, snow cover, sea/lake ice, pollution, and tropical storms. It is the best channel for monitoring volcanic plumes, aerosols, and dust storms.

<u>CHANNEL 2</u> records near-infrared radiation, with wavelengths of 0.73 to 1.1 microns, and is used primarily to discern clouds. Since liquid water absorbs near-infrared radiation more strongly than visible radiation, this channel is used for delineation of land-water boundaries. It can be used with channel 1 to locate areas of snow and ice melt. The NOAA AVHRR split channel allows comparison of the visible and near-infrared signals which aids interpretation of the data.

CHANNEL 3 records near-infrared radiation, with wavelengths of 3.55 to 3.93 microns. Channel 3 responds to emitted infrared and reflected solar energy. Emitted infrared is used normally only at night because of contamination by reflected solar energy during the day. Channel 3 is very sensitive to high energy sources and can be used to detect hot spots, such as gas flares, forest fires, volcanoes, and smoke stacks. Because of its warm temperature capability and low attenuation by water vapor, Channel 3 is excellent for thermal mapping of clouds and surface temperatures in the tropics where high moisture content often degrades longer wavelength channels. Channel 3's response to reflected solar radiation allows it to be used in the daytime to discriminate between pack ice and cloud, and ice and water.

CHANNEL 4 records infrared radiation, with wavelengths of 10.3 to 11.3 microns. Channel 4 is used both day and night for thermal mapping of cloud and sea surface. It is used for mid-latitude ocean currents, fronts, thermal studies, and cirrus cloud coverage. Channel 4 can be significantly degraded in the tropics by high moisture concentrations.

CHANNEL 5 records infrared radiation, with wavelengths of 11.5 to 12.5 microns. Channel 5 was designed to account for the effects of water vapor attenuation and to enhance determination of sea surface temperatures in the tropics.

2. Sensor Mechanics.

AVHRR sensors are mounted such that they are always oriented toward the earth. The key to the continuous operation of the radiometer is the scanning mirror, which rotates 360 times a minute. During each rotation, when the mirror is pointed at the earth, an imagery sampling process occurs which permits receipt of earth scene radiation at specified times during the scan. The radiation is then reflected from the scanning mirror through secondary optics which separates it into the discrete spectral bands. This results in 360 scan lines of imagery per minute and permits the very high resolution of 1.1 km at the satellite subpoint. The resolution degrades to 4 km at the edge of the image because the instantaneous field of view becomes progressively larger. The swath width, a function of spacecraft height, is about 25 degrees wide. Resolution is high and distortion is insignificant along a 10 degree swath. Images are produced by the combination of side to side crosstrack scanning and the forward motion of the spacecraft. Visual and infrared imagery are produced from the earth scene radiation simultaneously; the visual by silicon detectors and the infrared by metal segment detectors. The visual and infrared images are then digitized and transmitted to earth via S-band frequencies.

3. Secondary NOAA Sensors.

Secondary sensors on the NOAA satellite include a Tiros Operational Vertical Sounder (TOVS) that measures the vertical profile of temperature, water vapor, and ozone; a data collection and location system, and a Space Environment Monitor (SEM) that measures energetic particles. Secondary sensor data are not received at direct readout sites.

C. NOAA Satellite Imagery Dissemination.

There are three main Command and Data Acquisition (CDA) stations that receive NOAA imagery. These are located at Wallops Island, VA, Gilmore, AK, and Lannion, France. Reception from a NOAA satellite ceases at a CDA station when the satellite gets lower than 5° above the horizon, or a distance of 29° latitude from the CDA. Imagery from the NOAA satellite begins to be received by the user between five and ten minutes after the satellite leaves the CDA reception circle. It takes about 14 minutes to receive the complete horizon-to-horizon image from the CDA. Users can receive NOAA (polar orbiter) imagery by selecting the proper code on the Unifax II touchtone switching system.

II. Constraints on Meteorological Satellite Imagery Accuracy.

There are numerous factors during imagery sensing, processing, and dissemination that affect the quality and representativeness of the data. Some of these effects can be and are compensated for at various acquisition stages. Others cannot be corrected for, and their impact on the final photo product must be understood, although it frequently cannot be quantified.

A. Foreshortening.

This is a multifaceted phenomena which increases towards the edges of the imagery where resolution becomes poorer, distortion increases, and cloud location is displaced due to parallax error. Also, the amount of cloud coverage will be greater at the image edges because the sensor was actually looking into the sides of clouds. NOAA polar orbiter resolution varies from 1.1 km under the track of the satellite to 4 km at swath edges. For GOES satellites, resolution degradation is considerable because of the very broad field of view. For example, on a full disc GOES picture, the infrared resolution varies from 7 km at the subpoint to about 21 km at 60°N and 60°S latitude. Appropriate charts can be used to correct for cloud displacement. Resolution near the edge of imagery can be mapped, but not corrected for. <u>Cloud coverage errors cannot be corrected.</u>

B. Contamination.

Contamination occurs when radiation is sensed from multiple sources; for instance, thin cirrus over altocumulus. The result may be cirrus that appears warmer and lower, making it hard to recognize as cirrus. Contamination depends on cloud thickness, cloud spacing, sensor viewing angle, and the vertical temperature profile. <u>Contamination is not correctable.</u>

C. Attenuation.

Attenuation is a decrease in radiation received at the spacecraft due to energy scattering and absorption by the atmosphere. This attenuation results in clouds appearing colder and higher than they really are. The prime energy absorbers (water vapor, carbon dioxide, and ozone) can cause cloud top temperature readings up to 6°C too cold. The effect of attenuation increases as one moves from the subpoint to the edges of an image. The colder temperature sensed by the spacecraft causes the edges to have whiter gray shades. This is known as limb darkening. <u>Attenuation is not correctable.</u>

D. Temperature Averaging of the Infrared Sensor.

The GOES infrared sensor scans from west to east, producing an average gray shade (temperature) for pixels, or portions of the earth's or cloud surface that range from 7 km up to 21 km on a side. NOAA infrared sensors produce similar pixels from 1.1 km up to 4 km on a side. If we do not consider sensor lag, and if a cloud, small storm, or part of a storm does not completely cover the pixel, the cloud temperature, indicated by that pixel, will be an average of the actual cloud temperature and the temperatures of the surrounding clouds and ground. In other words, small, cold features are de-emphasized, and may be hidden. <u>Such averaging cannot be corrected for.</u>

E. Time Response Lag of the Infrared Sensor.

This problem is caused by an unavoidable time lag in sensor response as it scans across the earth scene. This results in convective type cloud tops being sensed further east and appearing warmer; thus suggesting cloud heights not as high as they may be. A GOES infrared sensor can only change its reading a maximum of 26°C per pixel. For example, if the surface temperature just west of a violent thunderstorm is 32°C, and a significant part of the cloud top near the west edge of the storm is -83°C, one pixel could indicate the correct 32°C; the next pixel indicating a temperature no lower than +6°C; the following -20°C; and the fourth -56°C. This time response lag produces errors, in this example, up to 0°C, +89°C, +63°C, and +27°C. <u>It is easy to see how response lag can render a small, tall, stationary, flash-flood-producing thunderstorm invisible, or vastly displaced on imagery.</u>

1-E-4

F. Miscellaneous.

At least six other factors, although perhaps not routinely significant, can contribute to difficulty in using imagery. Sensor lens/mirror aberrations, communication problems, receiving site imperfections, cloud-earth contrast, alignment of a sensed object in relation to the scan line, and sub-pixel sized cloud elements all can decrease the quality of a picture. Communications and receiving site imperfections can sometimes be corrected, the other factors cannot be compensated for.

Below is a summary of various constraints on imagery accuracy.

FACTOR	IMPACT	CORRECTABLE?	HOW?
Foreshortening	Cloud displacement	Yes	Use a correction chart.
	Resolution decrease	No	Resolution can be mapped.
	Cloud coverage error	No	---
Contamination	Clouds appear lower	No	---
Attenuation	Clouds appear higher	No	---
Pixel Averaging	Clouds appear lower	No	---
Infrared Sensor Lag	Clouds displaced east	No	---
	Clouds appear lower	No	---
Sensor Lens/Mirror Aberrations	Resolution decreased	No	---
Communications	Quality decreased	Maybe	Improve distribution
CDA Imperfections	Quality decreased	Yes	Maintainance and better quality control
Object/Scan Line Alignment	Detectability	No	---
Cloud/Earth Contrast	Surface/cloud confusion	Maybe	Enhancement Curves

Summaries from

CHARACTERISTICS OF WATER VAPOR IMAGERY

by Roger Weldon and Sue Holmes

SUMMARY, SECTION I.

Radiation intensity measured at the satellite, also referred to as "radiance", may be converted to brightness temperature or to image gray shade. Radiation arriving at the satellite in the infrared bands may be from the surface of the earth, from cloud tops, from water vapor, or some combination of the three. If little or no water vapor exists above, the brightness temperature will be representative of the cloud top or surface temperature.

If water vapor is present above the clouds, or if no clouds are present, energy will be reradiated by the water vapor which is usually at colder temperatures.

Radiation to the satellite from water vapor is emitted from some layer of finite vertical extent. The depth of the moist layer contributing radiation to the satellite is dependent upon the amount and density of the moisture existing in the radiation's path.

Several basic factors affecting the behavior of radiation may be considered. Although several of the factors usually vary in combination in the atmosphere, knowledge of their individual effects is useful to the understanding of water vapor imagery.

WAVELENGTH OF THE RADIATION

Of the operational "water vapor" channels on NOAA satellites, the 6.7 um and 7.3 um bands are most commonly used in image format. Of these, 6.7 um radiation is more highly absorbed by moisture. Therefore, smaller amounts of moisture, commonly at higher altitudes, are detected by 6.7 um radiation than by 7.3 um radiation. However, because of this, low level features are usually obscured on 6.7 um imagery.

THE AMOUNT OF WATER VAPOR IN THE PATH OF THE RADIATION

Increased amounts of moisture or water content in the radiation's path accomplish more absorption of the radiation arriving from lower levels. Therefore, if the air temperature decreases with height, larger amounts of moisture result in colder brightness temperatures.

THE VERTICAL LOCATION OF THE MOISTURE

   Since the vertical change of air temperature is commonly very large, moisture at high altitudes reradiates at low intensity contributing to colder brightness temperatures than at low altitudes. But, moisture in higher less dense air absorbs less radiation than the same amount at lower altitudes. Therefore, the brightness temperatures from moist layers at high altitudes are significantly warmer than the air temperatures at those altitudes.

THE TEMPERATURE OF THE AIR COLUMN

   The vertical change of air temperature, which is generally large, is necessary for the production of different brightness temperatures by varied moisture distributions. For the same atmospheric moisture conditions, warming or cooling of the temperature sounding will produce corresponding changes in the brightness temperature. Very cold air temperature soundings, such as those at high latitudes during winter on the cyclonic side of the jet stream, also have very small vertical changes of temperature. Therefore, changes of moisture cannot produce large changes in brightness temperature, or large variations of gray shade on images. Low latitude soundings, or summer season soundings in upper air anticyclones, have very large vertical temperature ranges. Very large gray shade variations can be produced by different moisture distributions.

   No valid simplified explanation exists for the relationship between atmospheric conditions and the variation of brightness temperatures, such as those displayed as gray shade patterns on water vapor imagery. The net intensity of the radiation arriving at the satellite is dependent upon the amount and vertical location of the moisture in combination with the air and surface temperatures.

   In the next section, moisture in varied layers shall be examined. The temperature sounding and wavelength will be held constant; but, the amount and vertical location of the moisture will vary in combinations.

SUMMARY, SECTION II.

   The satellite measures only the intensity of the radiation arriving from below. It does not detect the origin of the radiation. If the intensity is converted to brightness temperature, the relative vertical location of the emitting substance may be estimated. However, the same net radiation intensity at the satellite may be produced by various combinations of air temperature, humidity, and vertical distributions of moisture. This is particularly significant when moisture is distributed in single or multiple layers.

RANGE OF BRIGHTNESS TEMPERATURES

The range of brightness temperatures, computed or measured, is limited at the warm end by the skin temperature of the earth's surface; or, in some cases, by moisture or clouds at the top of a strong temperature inversion. With no middle or high clouds present, the coldest 6.7 um brightness temperatures computed or measured in our studies were near -40°C.

Under cloud free conditions, a practical range of brightness temperature for 6.7 um data extends from -40°C to +20°C. The range for 7.3 um data is smaller, (-20°C to +25°C) given mid or low latitude soundings. However, with dry, cold temperature soundings such as over continents in winter, 7.3 um data can become as cold as the 6.7 um data.

THE "CROSSOVER" EFFECT

When moisture is arranged in layers, the brightness temperature measured by the net radiation reaching the satellite is often unrepresentative of the air temperature of the layers. Layers of moisture at high altitudes produce relative warm brightness temperatures, since much of the high intensity radiation from lower warmer sources penetrates through the layer. Low level moist layers, which often absorb all or most of the upcoming radiation, produce warm brightness temperatures because the moist layer itself is warm. Because of the "crossover" effect between radiation penetrating from below, and radiation from the moist layer, the coldest brightness temperatures are produced by moist layers at middle altitudes. Since 6.7 um radiation is more easily absorbed by water vapor, its "crossover" altitude is higher than that of 7.3 um radiation. A 14 layer experiment was composed of moist layers with air temperatures and humidities commonly observed in the atmosphere. For such layers, the coldest 6.7 um brightness temperatures are most likely to be produced by moisture located near 500 mb. Those for 7.3 um radiation are most likely from moisture just above 700 mb.

THE "SENSITIVITY RANGE"

The net radiation intensity reaching the satellite in the 6.7 um band is very sensitive to differences in humidity within middle and high moist layers. The "sensitivity range" is a maximum near 400 mb. However, differences of water content in low moist layers - typically those below 700 mb - are not easily detected by 6.7 um radiation measurements. The "sensitivity range" of 7.3 um radiation is overall less than that of 6.7 um radiation; however, it remains sufficiently large at low levels to detect differences in humidity commonly found there.

MULTIPLE MOIST LAYERS

If no significant air temperature inversion is present, the addition of a moist layer either above - or below - an original moist layer will produce a colder brightness temperature. If a layer is added under an original moist layer with very large water content and density, such that all or most of the upcoming radiation is already absorbed; then, no appreciable reduction of brightness temperature will occur.

EXCERPTS FROM: SATELLITE INTERPRETATION
THE THIRD WEATHER WING/TECHNICAL NOTE - 81/001
EUGENE M. WEBER AND STEVEN WILDEROTTER 28 DECEMBER 1981

In this section a brief review of the characteristics of cloud patterns and types as shown in visible and IR photos will be discussed.

Cloud Patterns:

Synoptic events form cloud systems with various patterns. Some of the most common patterns are illustrated in Figures 14 to 17.

• Cloud Shield - A cloud shield is a broad cloud pattern which is not more than four times as long in one direction as it is wide in the other direction. A cloud shield (A) is shown across the Great Lakes and Ohio Valley.

• Cloud Band - A cloud band is a nearly continuous cloud formation with a distinct long axis where the ratio of length to width is at least 4 to 1 and the width is greater than 1 degree of latitude. A cloud band is shown at B in Figure 14.

• Cloud Line - A cloud line is a narrow cloud band in which the individual cells <u>are connected</u> and the lines are less than 1 degree latitude in width. Cloud lines are shown at C in Figure 15 and near D in Figure 16.

• Cloud Street - A cloud street is a narrow cloud band in which the individual cells <u>are not connected</u>. Several streets generally line up parallel to each other with each street not being more than 1 degree in latitude in width. Cloud streets are shown at E in Figure 15. See further cloud street discussion related to Figures 28 and 29.

• Cloud Finger - Cloud fingers develop on the forward edge of the frontal band and are often tied in a nearly-continuous fashion to the frontal clouds. These fingers generally extend in a more southerly direction than the frontal band. Figure 17 illustrates several cloud fingers, noted by the arrows, converging into a storm system located over the southeastern U.S. Cloud fingers generally end at the ridgeline.

• Cloud Element - A cloud element is the smallest cloud form which can be resolved in a satellite picture. Cloud elements are recognizable in Figure 15 within the areas noted by C and E.

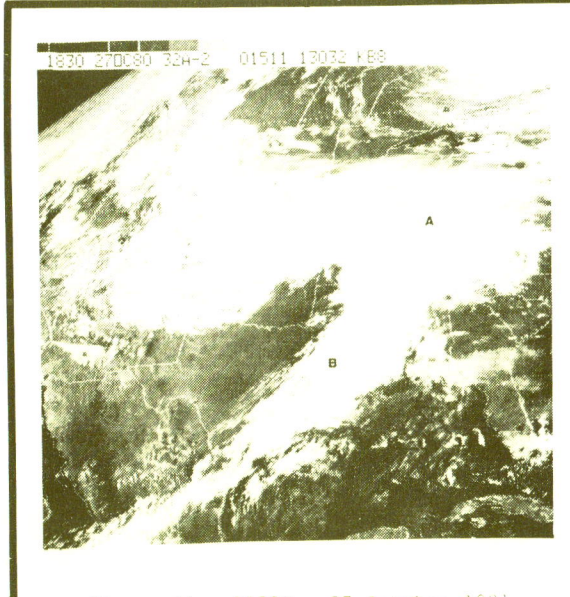

Figure 14: 1830Z 27 October 1981

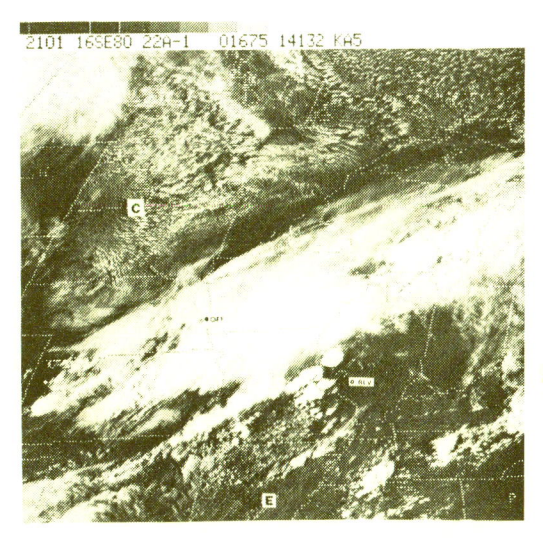

Figure 15: 2101Z 16 September 1981

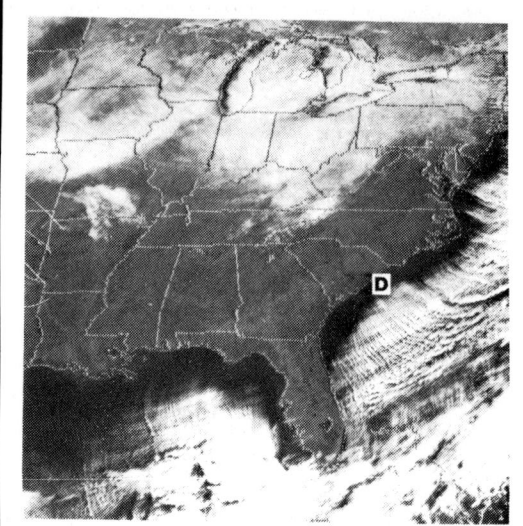

Figure 16:   1740Z   3 February 1981

Figure 17:   1930Z   10 February 1981

Figure 18:   2015Z   2 July 1981

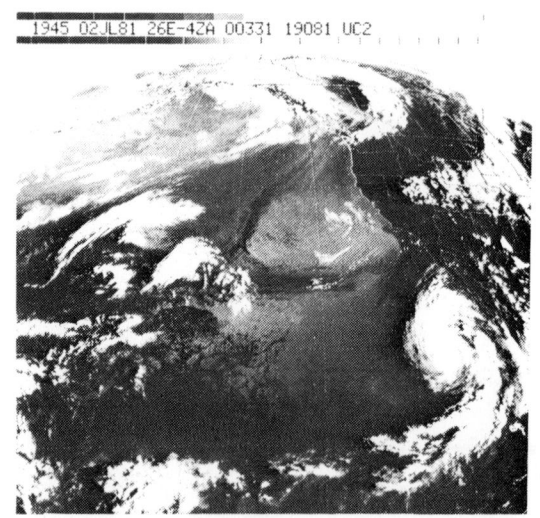

Figure 19:   1945Z   2 July 1981

Cloud Types:

Low Clouds

• Stratocumulus - Stratocumulus cloud fields appear as white, lumpy globular clouds in the visible (A and B in Figure 18) and as a consistent dull gray in infrared photos (if they can be seen at all; Figure 19). They are generally seen in closed cellular patterns with large numbers of relatively bright globular centers often connected to each other with darker, less dense clouds. As these cells continue to decrease in size, they take on a stratiform appearance and it becomes difficult to distinguish stratocumulus from stratus as shown off the California coast in Figure 20.

Figure 20:   1745Z   15 July 1981

1-G-2

Figure 21:   2030Z   22 April 1981

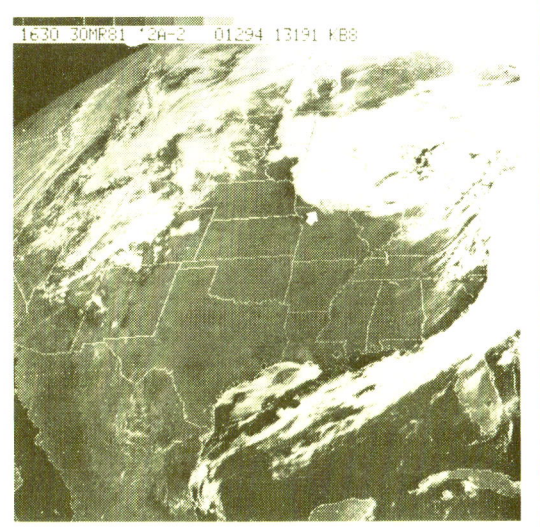

Figure 22:   1630Z   30 March 1981

Stratocumulus fields often form in three general locations in the northern hemisphere. The first is over water east of the subtropical anticyclone and south of the polar jet stream. An example was shown in Figure 18. The second is in the cold air advection area behind a cold frontal system as noted by the arrows in Figure 21. This is especially common off the east coasts of North America and Asia as the cold continental air encounters the warmer ocean currents. The third pattern is behind a comma cloud system where the low clouds just start to clear as indicated by the arrow in Figure 22. It is often difficult to distinguish cold air stratocumulus from stratus such as shown in Figure 22 because the cellular size is small and appears stratiform.

Figure 23:   2030Z   14 February 1981

Figure 24:   2001Z   14 February 1981

• Stratus - Stratus and fog both show a smooth flat texture in a visible picture as noted by the arrow in Figure 23. In the IR, stratus appears as a dull shade of gray (if they can be seen at all; Figure 24). Often, due to a temperature inversion during the cold season, the temperature of the land surface is colder than the temperature at the top of the cloud. If this occurs, it is nearly impossible to see stratus on an IR photo. Visible photos are not available during the night when many stratus events occur. Figure sequence 25 illustrates this stratus identification problem. In Figure 25a, a morning photo, a continental polar air mass prevails across the Midwest; the center is located in western Missouri. A stratus layer exists from Texas northward into Kansas, but it is very difficult to see because of the cold terrain. The eastern fringes of the stratus layer is barely discernible over eastern Texas as noted by the arrow in Figure 25a.

1-G-3

Five hours later, Figure 25b, the stratus tongue is more discernible across the Great Plains. A greater contrast between the cloud layer and the terrain evolve due to increase heating at the surface. Figure 25c, a visible, depicts the extent of the stratus layer that had existed during the night time hours.

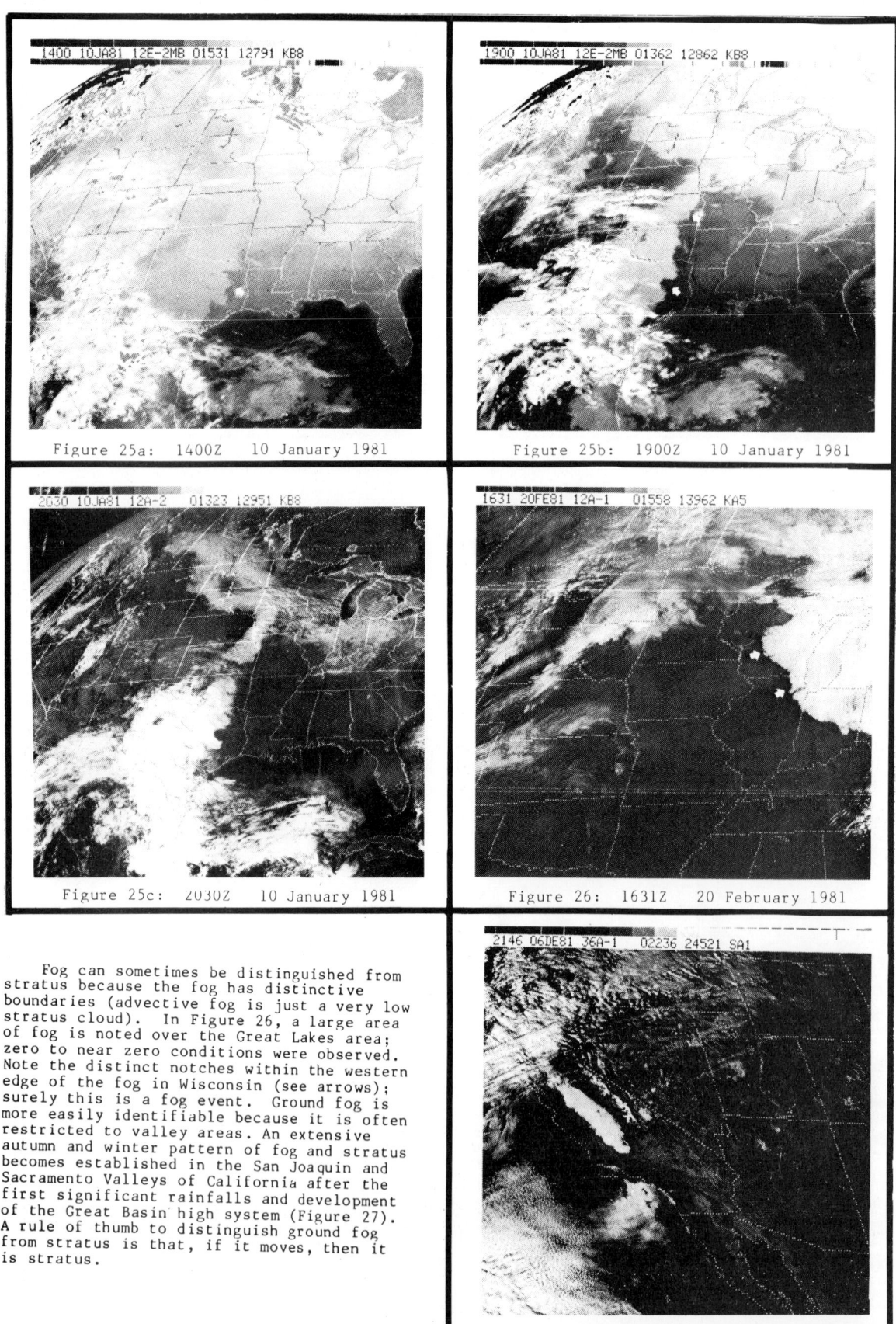

Figure 25a: 1400Z 10 January 1981

Figure 25b: 1900Z 10 January 1981

Figure 25c: 2030Z 10 January 1981

Figure 26: 1631Z 20 February 1981

Figure 27: 2146Z 6 December 1981

Fog can sometimes be distinguished from stratus because the fog has distinctive boundaries (advective fog is just a very low stratus cloud). In Figure 26, a large area of fog is noted over the Great Lakes area; zero to near zero conditions were observed. Note the distinct notches within the western edge of the fog in Wisconsin (see arrows); surely this is a fog event. Ground fog is more easily identifiable because it is often restricted to valley areas. An extensive autumn and winter pattern of fog and stratus becomes established in the San Joaquin and Sacramento Valleys of California after the first significant rainfalls and development of the Great Basin high system (Figure 27). A rule of thumb to distinguish ground fog from stratus is that, if it moves, then it is stratus.

• Cumulus - Cumulus and towering cumulus give a variable appearance in satellite photography depending on the degree of vertical development, the cloud slope and the solar angle of elevation. Cumulus clouds are found in bands, lines, streets and cellular patterns and are generally parallel to the flow. Cumulus congestus cloud bands may be parallel to either the flow or vertical shear.

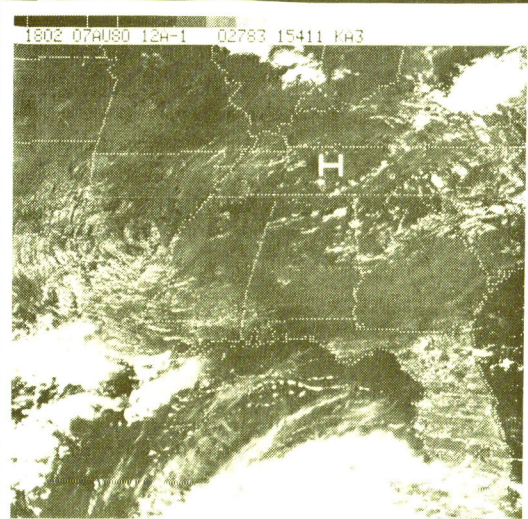

Figure 28: 1802Z   7 August 1981

During the summer convective season, when large areas of low-level moisture exist over land, circulation centers can be located on visible photos by following cumulus lines and streets. Two examples are shown in Figures 28 and 29. In Figure 28, an anticyclone is centered in central Tennessee; note the organized cumulus streets aligned with the low-level flow. In Figure 29 the remnants of a tropical storm can be located over central Texas by the organized streets of cumulus and cumulonimbus clouds.

During the cold season, cumulus with vertical development and stratocumulus lines and streets may be used as a guide to determine low-level flow within storm systems (see Figure 50).

A common event during the cold system is shown in Figure 30 over the Great Lakes and northeastern U.S. Stratocumulus and cumulus lines develop over and to the lee of the lakes as shown in Figure 30 when cold polar air moves across the warmer, moist Great Lakes.

Figure 30: 1847Z   8 December 1981

Figure 29: 1930Z   7 September 1981

Many cloud patterns in fields of cumulus clouds are associated with a maximum of cyclonic vorticity in the atmosphere. These cumulus fields are more easily identified over ocean areas associated with synoptic scale systems. Figure 31 shows a field of cumulus and cumulus congestus associated with vorticity over the northern plains and is noted by the arrow.

Middle Clouds.

• Altostratus and Altocumulus - Most altostratus and altocumulus clouds are difficult to depict in satellite photography due to their normal location within synoptic systems. When the middle cloud is not topped by cirroform clouds, they appear smooth in texture and are usually found in sheets. These characteristics make middle cloud layers hard to distinguish from cirroform layers in visible photos unless shadows can be seen. In Figures 32 and 33, bands of middle level cloudiness are recognizable due to the shadowing effect.

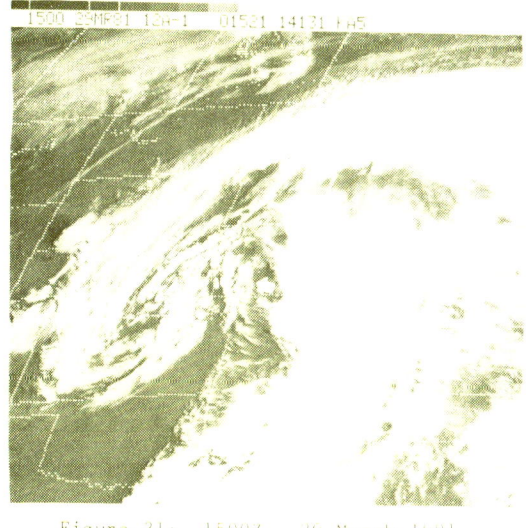

Figure 31: 1500Z   29 March 1981

1-G-5

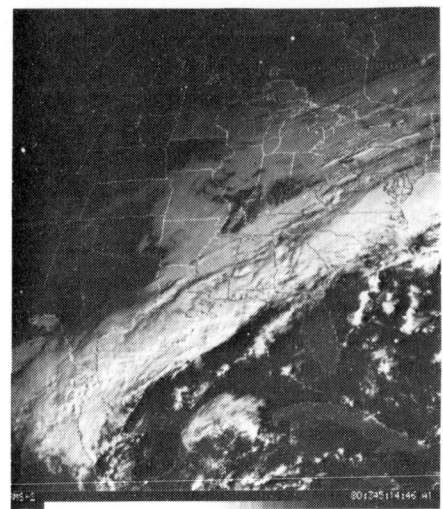

Figure 32: 1446Z 10 December 1981

Figure 33: 1916Z 18 December 1981

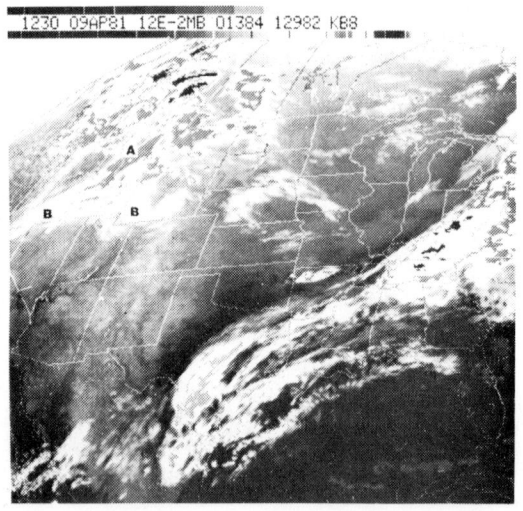

Figure 34: 1230Z 9 April 1981

In the IR, the middle cloud tops are seen as being lower than cirrus, so it is easy to distinguish altostratus from cirrostratus. Figure 34 shows an IR photo with cirrostratus at A and altostratus at point B. A middle cloud deck which is usually easily detectable in the visible is the lee-side mountain wave lenticular clouds (see Figure 43, page 12). These clouds are indicative of mountain wave turbulence.

High Clouds:

• Cirrus - Thin cirrus appears wispy and fibrous in the visible with the possibility of the terrain being seen through the cloud (see C in Figure 35). Thin cirrus is very much affected by contamination. Sometimes the cloud is so pronounced in the IR that it is mistaken for an indication of major cyclogenesis (see Figures 6 and 7, page 3). Figure 36 shows the same region of cirrus as Figure 35, but one-half hour later.

Figure 35: 2145Z 17 April 1981

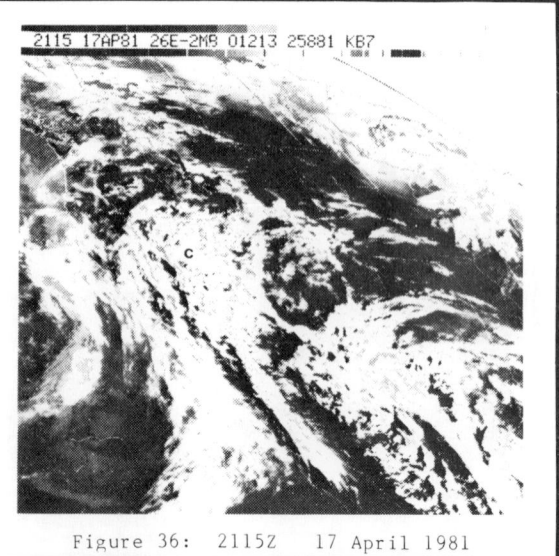

Figure 36: 2115Z 17 April 1981

• Cirrostratus - Cirrostratus is usually thicker than wispy cirrus. Cirrostratus occurs in extensive sheets. In the visible, the cloud appears to be a uniform, smooth white while in the IR various shades of grays and blacks reveal colder, higher tops. In Figures 37 and 38, visible and IR respectively, a cirrostratus layer associated with the jet stream stretches southwest to northeast across Mexico and Texas. In both figures, short cirrus filaments or streaks extend southward from the cirrostratus layer across Texas and Mexico as noted by the arrow. These filaments are observed frequently equatorward and on the anticyclonic side of the jet stream. They are most likely associated with vertical shear, and as such, are unreliable indicators of the upper level wind flow. They generally terminate at sharp upper ridgelines, however, they can carry across ridgelines of lesser amplitude. In the enhancement IR photo, Figure 38, note the exaggeration of the filaments.

Figure 37: 1930Z 25 January 1981

Figure 38: 2000Z 25 January 1981

Figure 39: 2032Z 3 June 1981

Figure 40: 2130Z 4 June 1981

Figure 41: 2132Z 10 July 1981

• Anvil Cirrus - Anvil cirrus is blowoff from thunderstorms. Figure 39 depicts several groups of thunderstorms across the southern plains, southern Rockies and Mexico. Anvil cirrus has a very sharp upwind edge (A), and a fuzzy downwind edge (B). The anvil cirrus points out either the direction of the upper winds or the direction of the vertical shear. In Figure 39 it is apparent over Mexico (C) that the upper winds over this area are from a southwest direction. Often the blowoff from numerous thunderstorms combine to form an extensive cirrostratus shield as seen at D in Figure 39.

A second example is shown in Figure 40 where anvil tops can be used to determine the direction of the upper winds. This is shown by the long arrows in Figure 40.

• Cumulonimbus - Cumulonimbus cells may form into patterns similar to the smaller convection clouds. Each cell is circular in shape as noted across the Gulf coastal states in Figure 41. They appear white in the visible and may cast shadows on lower cloud decks. To the east of an advancing squall line, a cirrostratus shield may develop due to excessive anvil blowoff. Not all cumulonimbus cells have anvils, and their IR temperatures are dependent upon the height of the cell.

Turbulence:

Satellite photos give some signs in areas where turbulence should be expected. Some of these telltale cloud patterns are:

• Turbulence of some degree can be expected near any convective activity. Greater amounts of turbulence occur in the enhanced cumulus pattern near vorticity centers or in the dry slot of the comma cloud. Figure 42 exemplifies enhanced cumulus at location D and a dry slot at E.

• Turbulence occurs near jet streams. Around the polar jet, turbulence can be expected in the upper two-thirds of the cirrus shield. Around the subtropical jet turbulence can be expected in the lower two-thirds of the cirrus shield. Stronger turbulence should be expected wherever the jet stream turns. Near the turn, turbulence will extend a little north of the cirrus shield.

• Mountain-wave turbulence is indicated by lee-side rotor or lenticular clouds. Figures 43 through 45 depict mountain wave activity respectively over the Rocky Mountains, Black Hills, and Appalachian Mountains.

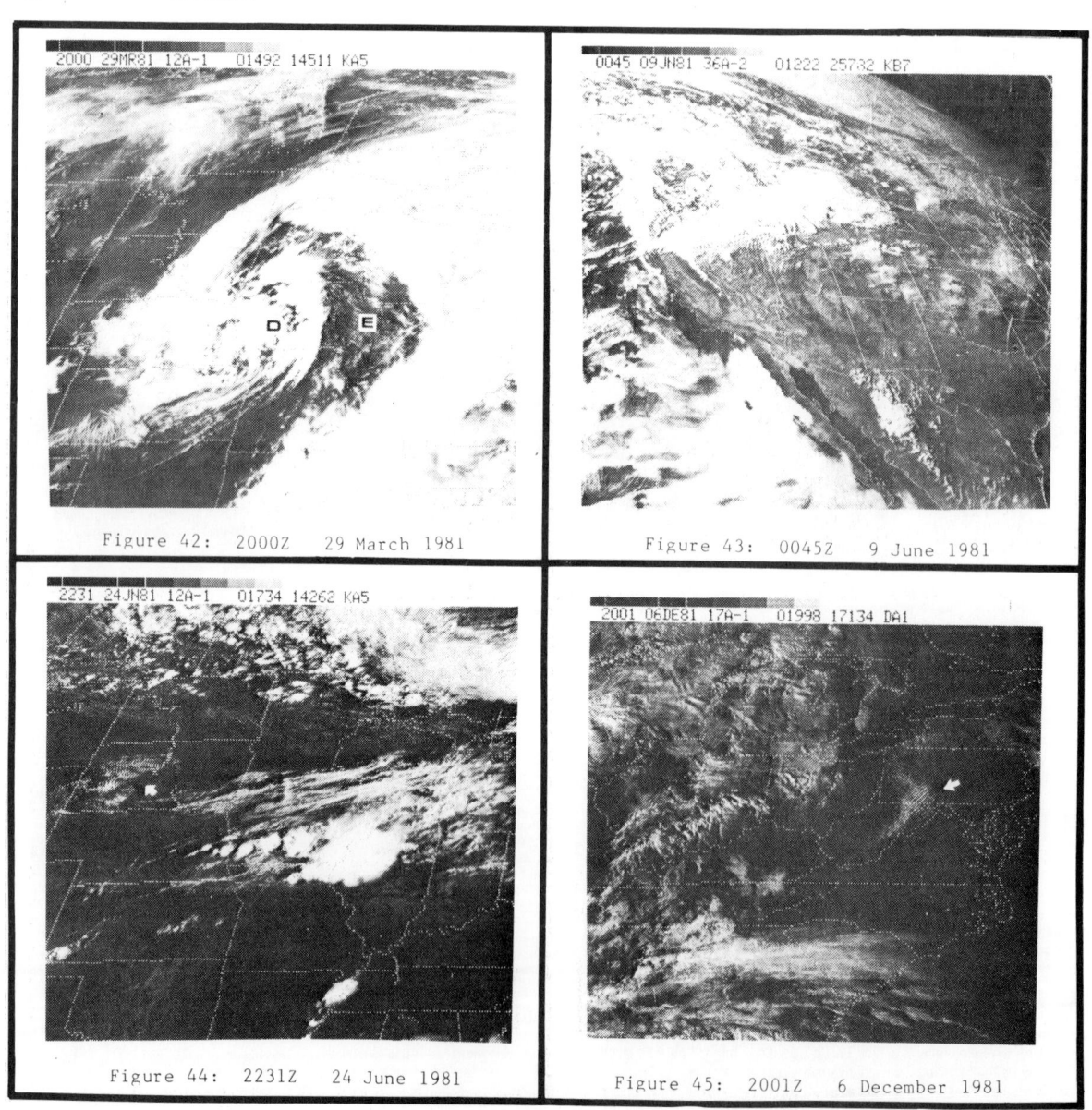

Figure 42: 2000Z   29 March 1981

Figure 43: 0045Z   9 June 1981

Figure 44: 2231Z   24 June 1981

Figure 45: 2001Z   6 December 1981

Ice and Snow Cover:

During the cold season, forecasters may find it difficult to interpret clouds from ice and snow cover on satellite photos. Snow areas are smooth, except for dendritic patterns which outline mountains and valleys, while clouds generally have a more textured appearance. In Figure 46, the arrows point to three locations of mountainous snow cover. Some scattered cloudiness is shown over the northern Rockies which makes it difficult to identify snow cover areas.

Over relatively flat terrain, snow cover can best be identified by geographical features such as unfrozen rivers and lakes located within the snow area. Additionally, clouds tend to move while the snow cover will appear unchanged over a short period. Over the Great Plains, snow swaths are generally recognizable after storm passage because of the absence of low clouds (dry continental polar air). Figures 47 and 48 illustrate these points. Figure 47 depicts a storm system over the Great Plains; twenty-four hours later, Figure 48, a snow swath (4" to 6") can be seen across northeastern Colorado and southern Nebraska. The cloud area associated with the receding storm system has moved away from the snow swath.

Snow cover areas across the Great Lakes - Ohio Valley eastward to the East Coast are sometimes difficult to identify on satellite photos because of persistent cloudiness (mostly cold air stratocumulus and stratus). Figure 49 reveals snow cover and some low cloudiness across the northeastern U.S.; two of the finger lakes in New York can be seen in the snow cover area.

Some factors affecting the appearance of snow cover will be the snow depth, age, terrain over which it fell, the cloud cover, vegetation, the solar angle, and whether or not it has rained since the snow fell. Generally, the brighter the snow, the deeper it is; beyond a depth of four inches this rule is no longer valid. The older the snow, the darker it will appear, especially if it has rained since it fell. The shade of snow cover will also depend on the vegetation growing in the area. Snow in a forested area looks milky gray. Ice can sometimes be distinguished from snow by its texture and location.

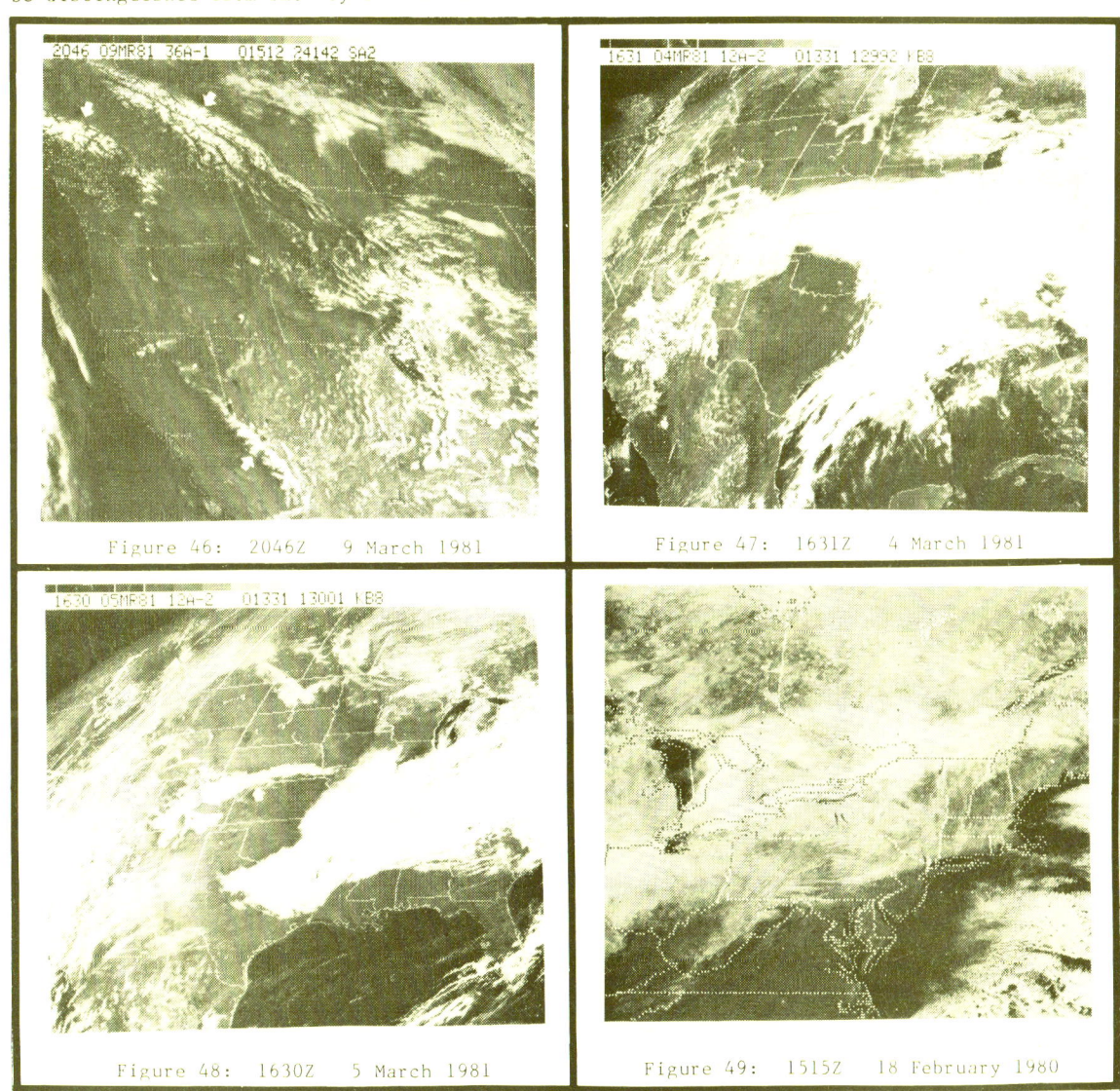

Figure 46: 2046Z 9 March 1981

Figure 47: 1631Z 4 March 1981

Figure 48: 1630Z 5 March 1981

Figure 49: 1515Z 18 February 1980

Surface/Lower Level Winds

Wind Direction:

Several rules of thumb may be used to estimate surface and lower level wind direction and speed. The direction may be found by cumulus or stratocumulus lines; these line patterns are often associated with extratropical cyclones as shown in Figure 50. In Figure 50, note the change in cloud alignment (indicated by the arrows) east and west of the vorticity comma tail noted at point B. The wind direction is usually parallel to these lines.

Wind Speed:

Wind speeds can be estimated to a limited degree by the appearance of closed or open cellular patterns. Open cellular cloud patterns are found in areas of cyclonic flow and closed cellular patterns occur most often in anticyclonic flow. Figure 51 illustrates these two cellular patterns off the eastern Seaboard and Gulf Coastal states. At points A and B, open cellular clouds are noted indicating low-level, cold-air cyclonic flow (vertical mixing and turbulence). The pattern changes to closed cells in the area of C which indicates low-level anticyclonic flow (stablization). At point D, the lower cloud layer changes characteristic from stratocumulus over the eastern Gulf of Mexico to stratus over eastern Mexico (warm moisture advection). Figures 52 and 53 respectively depict the surface and gradient level wind charts four hours prior to the satellite photo shown in Figure 51. In Figure 53, note where the divergent asympote lies; the transition from open (B) to closed (C) cellular patterns over the Gulf approximates the divergent asympote.

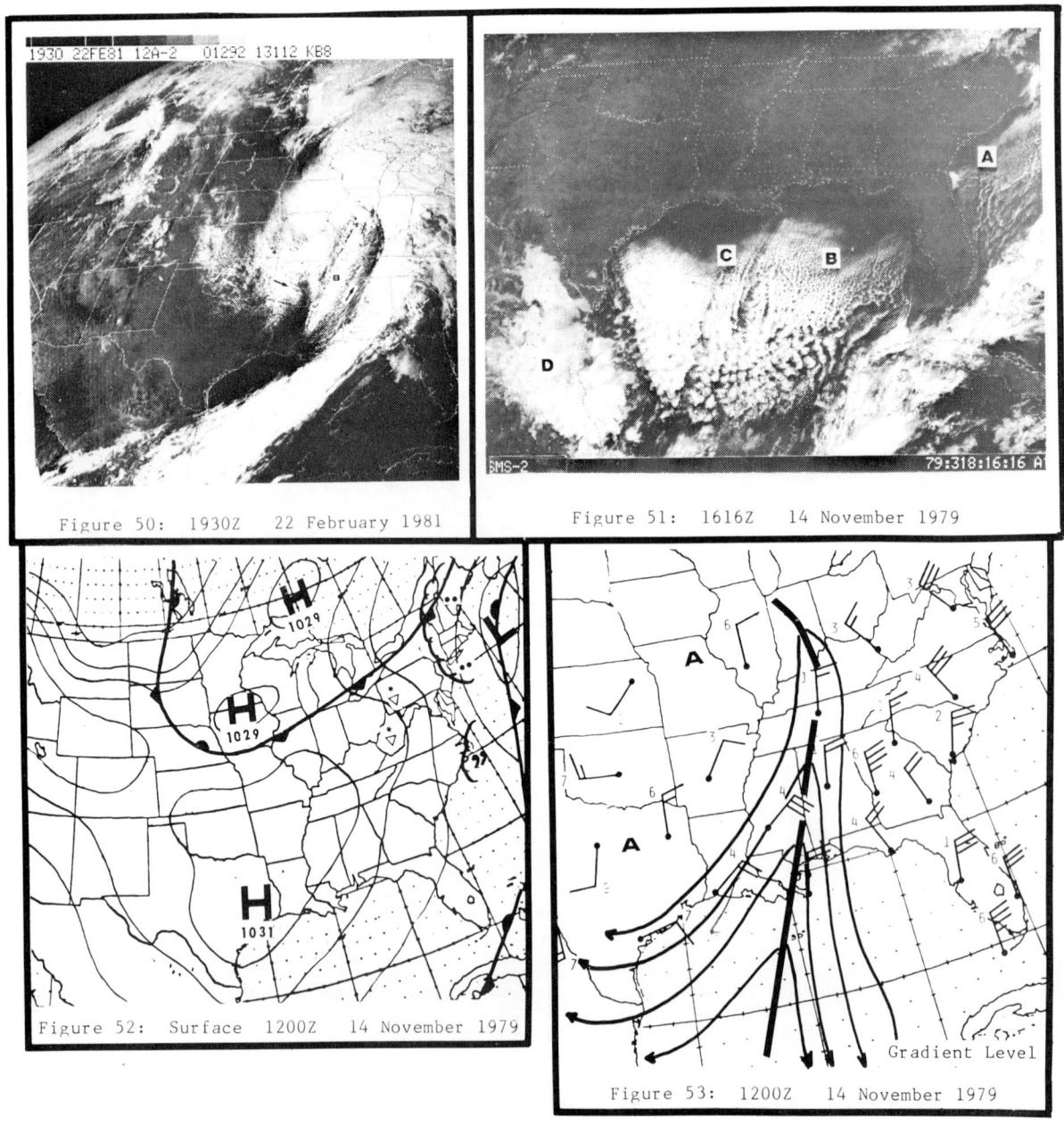

Figure 50: 1930Z 22 February 1981

Figure 51: 1616Z 14 November 1979

Figure 52: Surface 1200Z 14 November 1979

Figure 53: 1200Z 14 November 1979

Stratocumulus - If there are only isolated stratocumulus clouds, not in a linear pattern, the winds are usually less than ten knots. If the stratocumulus are in a linear pattern, the wind speed is normally between ten and twenty knots.

Cumulus - Open-cell cumulus cloud shapes are even better for determining surface and lower-level wind speeds. Figure 54 depicts various stages of open-cell cumulus with respect to wind speeds. If the cells are circular shape, then the winds are less than ten knots (Figure 54a). When they start to become oval shape, the winds are between ten and twenty knots (Figure 54b). When the cells become horseshoe shaped, the surface winds are between twenty and thirty knots (Figure 54c). Finally, when the cells become severely elongated (Figure 54d), the winds are greater than thirty knots. In Figure 55, an enhanced IR, horseshoe-shaped, open-cell cumulus are recognizable within the cold, cyclonic flow of a Great Plains storm system (see related visible photo, Figure 50).

Mountain-Wave Clouds - Mountain-wave clouds may also be used to determine the surface wind speed. The formula is:

$V = 6WL + 12$

WL is the cloud wavelength in miles.
V is the wind velocity in miles per hour.

If there is a series of terrain ridges upstream from the wave, this technique becomes difficult to apply due to overlapping waves.

Anomalous Lines - Anomalous cloud trails or lines can be seen in persistent marine anticyclones where ship exhaust gases have seeded the low-level stratocumulus and caused an overcast trail (Figure 56). These trails will persist sometimes for days in areas of light winds.

Upper Level Winds:

Upper winds can be estimated from satellite pictures using the following methods:

• Size of Anvil Cirrus - If the anvil continues to remain about the same size as the parent cell, then the upper winds are light. If the anvil is about double the size of the parent cell, the wind speed is approximately twenty-knots. If the anvil is triple the size of the parent cell, the wind speed is about thirty knots. To continue, multiply the number of parent cells which can fit into the anvil blowoff by ten to obtain the upper wind speed.

• Transverse Cloud Bands - Transverse lines or bands are sometimes observed in jet stream cirrus, especially during the cold season. These cloud patterns appear as small-scale lines at an angle almost perpendicular to the main jet or maximum wind flow and they indicate horizontal divergence away from the wind core. These lines look somewhat like waves, but are much more irregular than mountain waves and are observed over oceans as well as over land. Figures 57 (visible) and 58 (IR) depict transverse lines and cirrus filaments in jet stream cirrus across the southern U.S. The jet stream is generally to the left of the lines and generally one degree to the north of the northern edge of the cirrus shield as shown in Figure 58. Speeds in the transverse bands are usually 80 to 100 knots.

Note: Transverse banded cirrus is specifically associated with wind speed maximums and occurs in close proximity to the jet core. Transverse banding is shown near location A in Figure 57. In contrast, cirrus filaments, as noted at locations B and C in Figure 57, may occur equatorward of the jet core, but do not yield specific information about the winds. A more specific example of wind maximum associated with transverse banding will be shown in Section 1, Part II (Figures 82 through 87).

• Billow Cirrus - Billow cirrus is formed and maintained by wave motion just as lee waves are. They also have a "washboard" appearance. However, they are not associated with terrain influences. They are bands of very thin cirrus. The winds are usually perpendicular to the clouds much like transverse lines. The formula for wind speed using billow clouds is:

$V = 18.4WL + 16.4$

WL is the wavelength in nautical miles.
V is the velocity in knots.

The largest problem associated in using billow clouds to determine jet stream speed is that they are so thin and lower clouds may mask the cloud wavelength.

NOTE: Billow clouds may appear in the lower levels of the atmosphere.

• Cirrus Streaks - Cirrus streaks form parallel to the flow and to the right of the jet stream when there is insufficient moisture to form a full cirrus shield. They are usually observed on the backside of troughs. Winds of greater than sixty knots are usually required for their formation. In Figure 59, cirrus streaks stretch across the southern U.S. and the Gulf of Mexico.

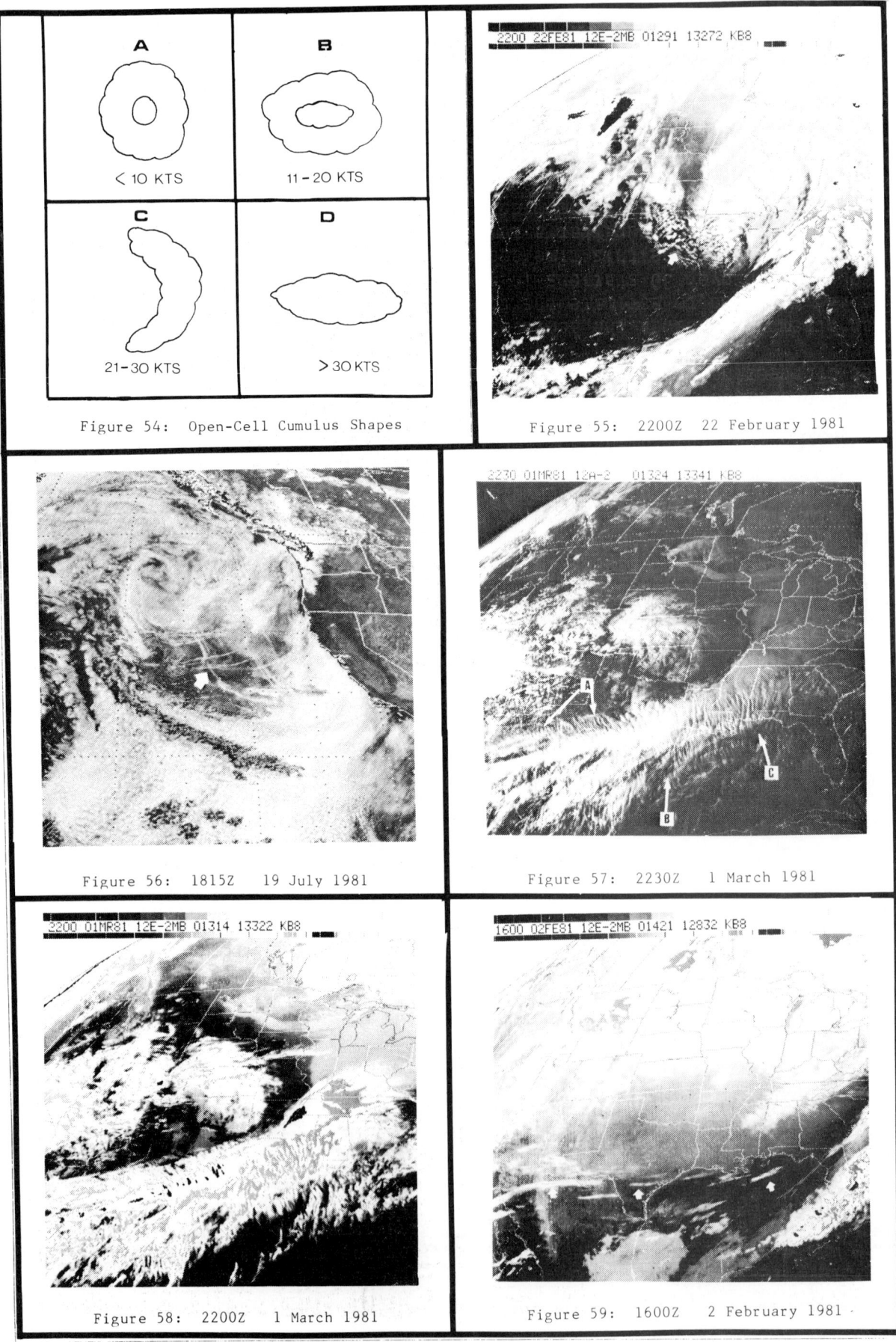

Figure 54: Open-Cell Cumulus Shapes

Figure 55: 2200Z 22 February 1981

Figure 56: 1815Z 19 July 1981

Figure 57: 2230Z 1 March 1981

Figure 58: 2200Z 1 March 1981

Figure 59: 1600Z 2 February 1981

Striations/Shadows:

   Striations are narrow, straight or curved streaks in clouds which appear in an overcast area. Striations frequently give indications of the orientation of the thickness lines. A low sun angle causes highlights and shadows which produce the striated pattern. Figures 60 and 61 depict morning visible photos which highlight striations. In Figure 60, note the cyclonic striations over the Great Lakes area as noted by the arrows.

   A second example is shown in Figure 61. Along with striations, shadows are also noticeable which gives the Great Plains storm system an effect of depth. Several cloud systems, marked by separate shadows, are as indicated by A, B, and D. The highest layer is shown at A (baroclinic zone cloud shield), a lower cloud layer (middle clouds) is shown at B (vorticity comma), and the lowest layer (low clouds) appears at location D in Kansas. East-west striations across Nebraska and Iowa (area C; north of the low clouds) reflect the thickness pattern (deformation zone) within the occluded segment of the frontal system.

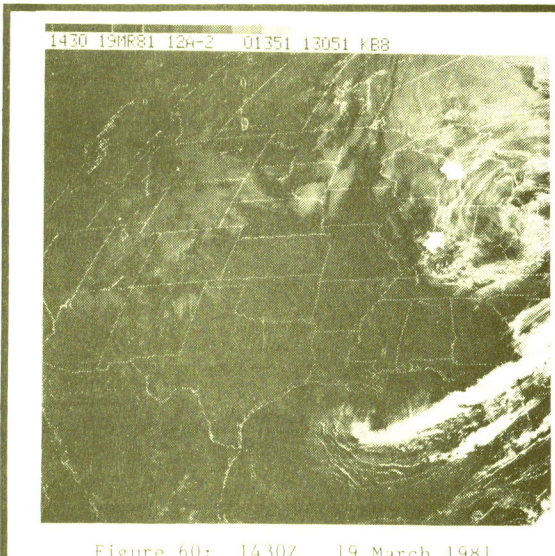
Figure 60: 1430Z   19 March 1981

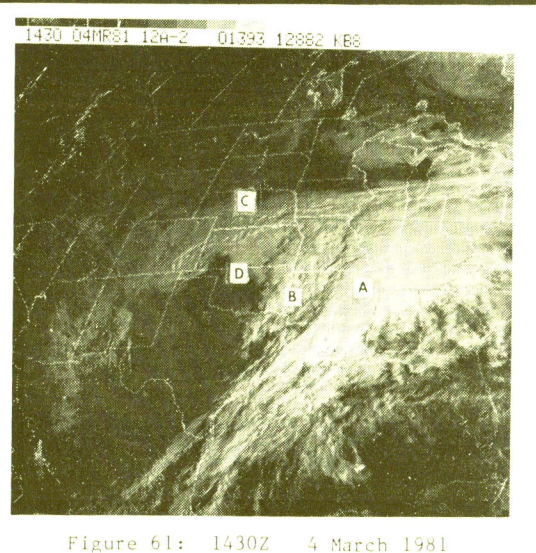
Figure 61: 1430Z   4 March 1981

Lithometeors:

   Wind-driven particles such as dust, sand, smoke and volcanic ash may dim or obliterate surface features. They are characterized by a dull, hazy, filmy appearance similar to thin cirrostratus in visible photos; IR photo may or may not reveal these lithometeors.

   • A duststorm event associated with a Great Plains storm system is shown over northwestern Texas in Figure 62. Surface wind directions were from a westerly component; wind speeds were reported between 45 and 50 knots. In the IR, Figure 63, the dust area is apparent and is within the same gray shade scale as low clouds located over the northern Texas panhandle and in the Gulf of Mexico.

   • A large area of smoke from Canadian forest fires is shown from southern Canada across the upper Midwest to the Great Lakes - Ohio Valley area as indicated by the arrows in Figure 64. Notice how the smoke layer across the central Midwest flows into the Rocky Mountain disturbance; the smoke layer has the same inflow pattern that a cloud layer would have if present.

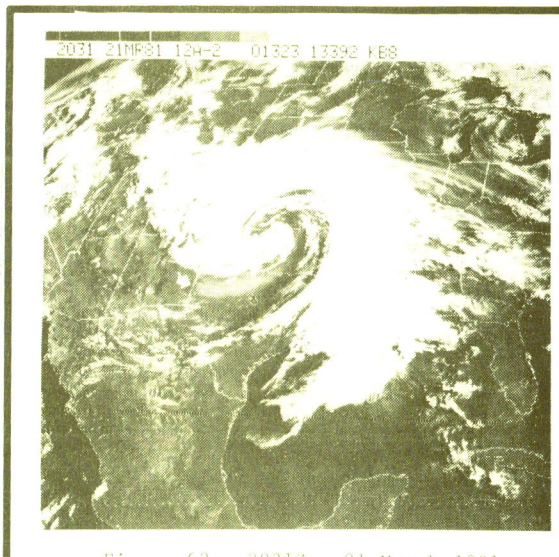
Figure 62: 2031Z   21 March 1981

Figure 63: 2100Z   21 March 1981

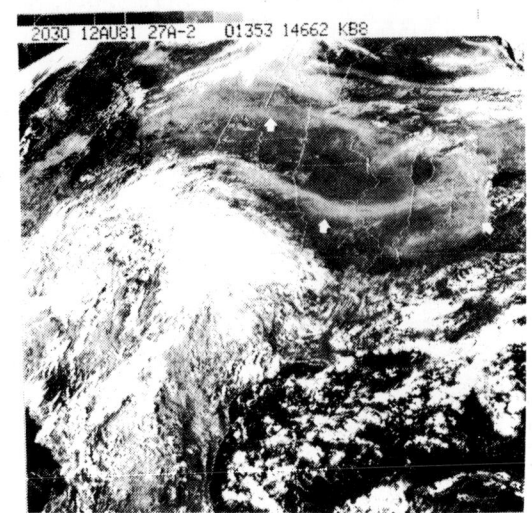

Figure 64:   2030Z   12 August 1981

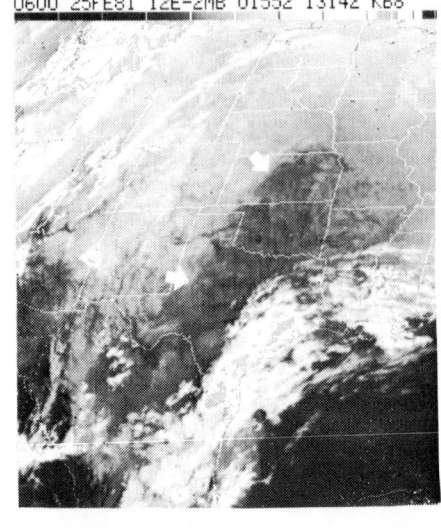

Figure 65a: 0600Z   25 February 1981

Low-Level Moisture Areas:

Enhanced IR satellite pictures can provide forecasters with an additional tool for forecasting the extent of warm air advection stratus and fog at night. Frequently, areas where fog and stratus are more likely to form over or advect into will appear as relatively dark areas on enhanced satellite pictures (after Gurka, 1976) (1). These dark areas are normally found downwind from a moisture source, such as the Gulf of Mexico and appear to outline the boundary of relatively moist air in the lower levels. This boundary can be monitored and used as an estimate of the extent of nocturnal fog and stratus developement and advection. During the night, moist air absorbs the earth's radiation, becomes warmer, and reradiates this energy in all directions - part of it back towards earth. In areas not covered with moist air, the earth's radiation is loss to space. The net effect is a warmer earth in areas covered by air with high moisture content which appears darker on infrared photographs.

Figure sequence 65 shows this event. In Figure 65a, an IR, a dark area can be seen across the southern and central plains; the western boundary of the moisture area is noted by the arrows. This dark area represents a tongue of moisture advection although stratus is not evident. Stratus is occurring across eastern Texas and offshore, but nearly all of the stratus is hidden by a cirrus layer. The northern limits of the dark area extends well to the north reaching into southern Nebraska. From this IR photo, the potential for stratus formation or advection would be forecast as far north as Nebraska provided no air mass changes would occur.

The visible imagery, ten hours later, Figure 65b, depicts subsequent stratus advection. In Figure 65c, (24-hours later from Figure 64b and 34 hours later from Figure 65a) stratus has indeed, advected as far north as Nebraska and has reached into the Dakotas.

Of course, the extent of the moist air at the surface can be located with surface dew points. The satellite, however, has much higher resolution than the surface observation network.

Figure 65b: 1631Z   25 February 1981

Figure 65c: 1630Z   26 February 1981

Terrain Features:

Forecasters should not be confused in interpreting terrain as low clouds on IR photos. In Figure 66, a large grey area is shown over the Rocky Mountains (location G; a small patch of cirrus is shown over the Front Range); the gray shade approximates a layer of low clouds. The related visible photo, Figure 67, reveals that nearly all of the Rocky Mountain areas are cloudless.

A second gray shade area, which could be misleading to novice satellite interpreters as low clouds, is faintly shown across the area noted by H in Figure 66 (arrows mark the western boundary). This grey cloud-free area is cold air; it closely follows the cyclonic configuration of the cold air stratus pattern, location J, associated with the Great Lakes storm system.

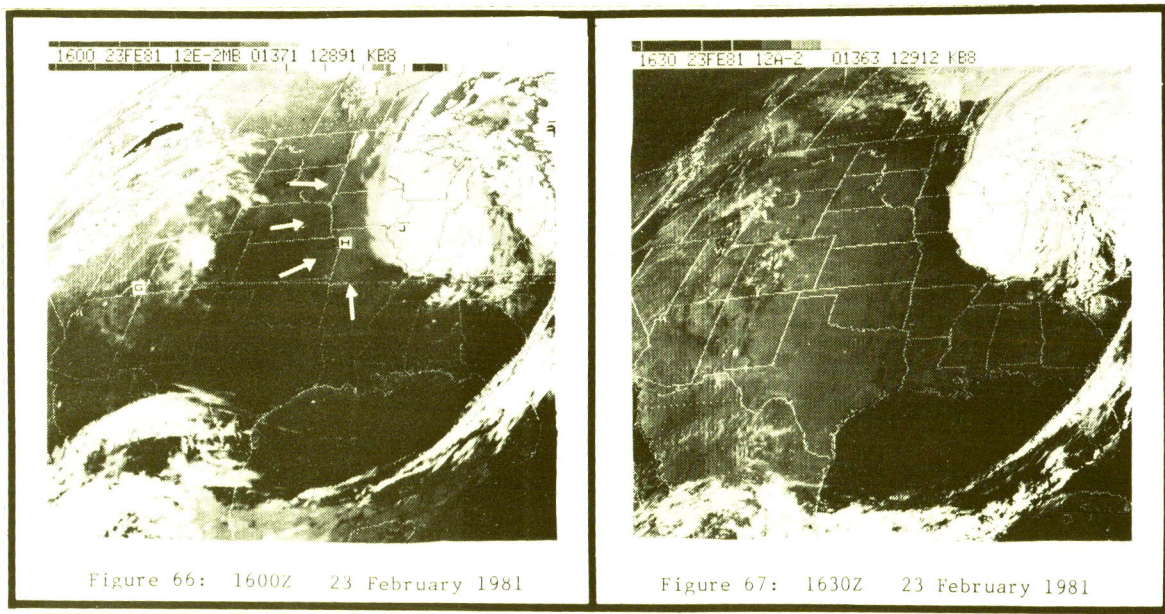

Figure 66: 1600Z   23 February 1981   Figure 67: 1630Z   23 February 1981

Satellite Photos Courtesy of:
  AFGWC
  SAC/WSU
  DET 1, 9WS

Figures 274, 275, 276 and 277 courtesy of
National Environmental Satellite Service
(NESS).

All conventional analyses shown throughout this Technical Note are duplicates of either NWS's Daily Weather Maps Weekly Series or facsimile charts.

U.S. DEPARTMENT OF COMMERCE

National Weather Service/National Environmental Satellite Service
SATELLITE APPLICATIONS INFORMATION NOTE 76/25

THE USE OF THE HB IR ENHANCEMENT CURVE TO MONITOR
THE ONSET OF FROST CONDITIONS IN THE WESTERN STATES

James J. Gurka and Ralph K. Anderson
NESS, Applications Group

The Hb enhancement curve (Fig. 1), which is used operationally in the Western Region, has a black contour that covers the temperature range from 0°C (32°F) to -3.3°C (26°F). Theoretically, on clear nights, this contour could be used as an indicator of the ground surface reaching freezing and the onset of frost. However, there are no ground temperature data available to check the accuracy of the satellite-observed temperatures on a given evening. The question then arises as to whether the relationship between satellite-observed ground temperatures and the six-foot instrument shelter temperatures is consistent enough to calibrate the temperature values associated with the enhanced infrared (EIR) contouring. Satellite data could then be used to indicate the onset of freezing conditions in remote areas.

In an attempt to answer the above question, a grid with station locations (Fig. 2) was used to estimate the time at which the edges of the freezing contour passed over 30 stations in nine Intermountain states. It is important to determine this time accurately since it is only at the "breakpoint" of a contour that a temperature is defined. For example, if a station is located in the middle of the black contoured area on the Hb curve, the temperature could be any value between 0°C and -3.3°C. However, at the instant that the contour spreads over a station, the ground temperature should be 0°C.

On the evening of October 18-19 of this year, in the Western Region, the colder edge (C on Figs. 1 and 2) of the freezing contour on the Hb curve corresponded fairly closely to an instrument shelter temperature of 1°C (34°F). Thus in this case, the average difference between the satellite-sensed group temperature and the six-foot shelter temperature was 4.4°C (8°F). Figure 3 is a plot of the time that the instrument shelter temperature dropped to 1°C versus the time that the colder edge of the black contour spread over the station.

The best fit straight line on Fig. 3 is Y = .86X + 1.16 with a correlation coefficient of .96. The average error from the best fit straight line is 47 minutes with 19 out of the 30 cases having an error less than the average. The larger errors, such as the one case with 2.62 hrs. and another with 2.44 hrs., occurred when the contour remained stationary near a station location for several pictures, and thus small gridding errors could have made substantial differences. The theoretical best fit straight line should be Y = X. The average error from Y = X on this evening was 52 minutes, with 20 out of the 30 cases having an error less than 45 minutes. Therefore, there appears to be a close and consistent relationship between the satellite-observed temperature and the instrument shelter temperature.

There were several limiting factors in this study: (1) only hourly temperature reports were available; (2) the enhancement IR (EIR) pictures were taken hourly, and thus the arrival of the gray contour and the time at which the termperature dropped to 1°C were interpolated; (3) the resolution of the IR sensor is 8 km, so that the imagery presents a ground temperature that is an integrated radiational field from the ground, tree tops and other vegetation; and (4) slight gridding errors could have made a substantial difference for estimating the temperature at a point. Another possible source of error could be the amount of moisture in the atmosphere. The magnitude of error in satellite-derived temperatures due to water vapor absorption as a function of precipitable water content is shown on Figure 4. The precipitable water chart for 1200 GMT on the 19th (Fig. 5) clearly shows that atmospheric moisture was minimal in this case.

In order to use the EIR satellite pictures to monitor the spread of freezing temperatures, the user should: (1) be familiar with what instrument shelter temperature corresponds to the onset of frost in the area of interest; (2) determine what instrument shelter temperatures are associated with the edges of the black contour in cloudless areas as early in the evening as possible; (3) estimate what portions of the contour will be associated with the onset of frost in the agricultural area of concern.

On the basis of this one case, there appears to be a fairly consistent relationship between satellite-derived temperature and instrument shelter temperature. Therefore, knowing what that relationship is on a given evening and what temperature is associated with the onset of frost in a particular location, the EIR satellite pictures, and in particular the Hb curve, could be used to estimate the spread of frost conditions.

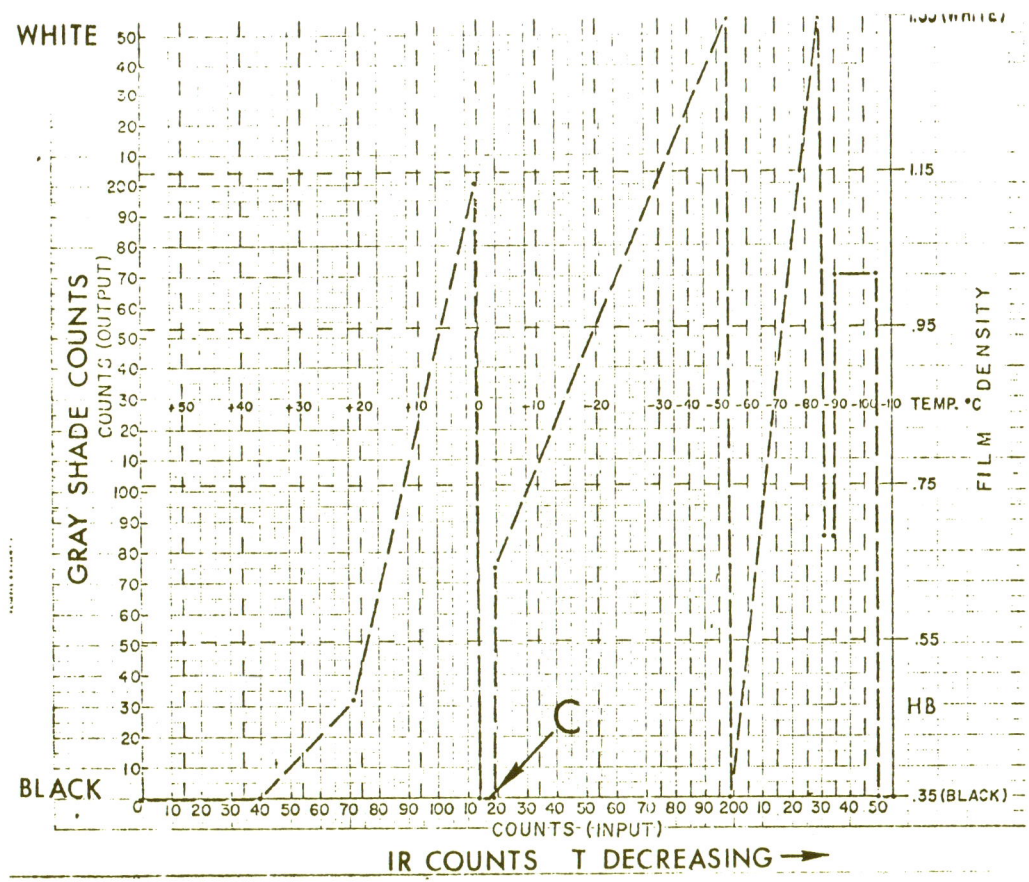

Fig 1. Hb IR enhancement curve: gray shade vs. temperature

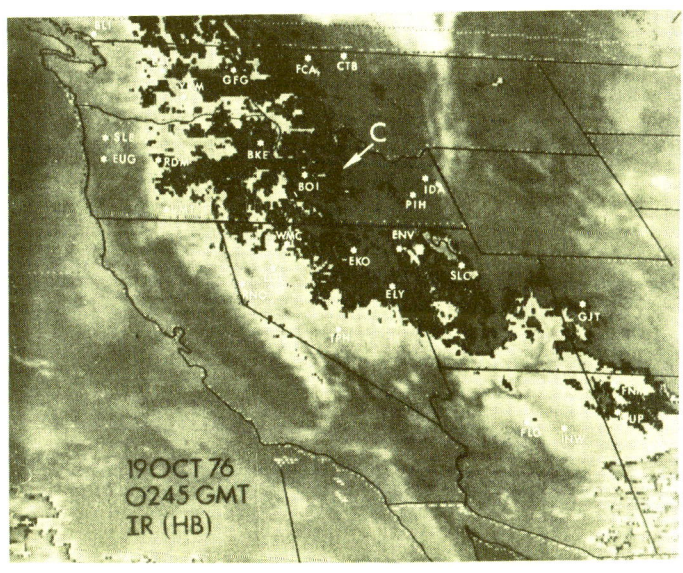

Fig 2. SMS-2 8-km IR picture with station locations, 19 Oct 76, 0245 GMT

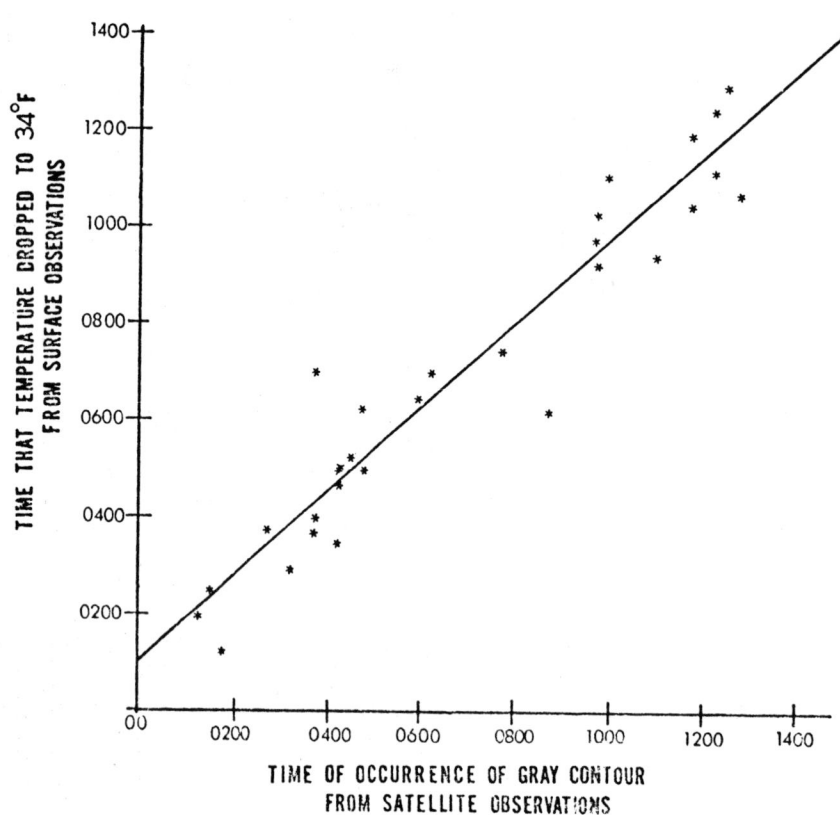

Fig 3. Time that temperature dropped to 1°C as a function of time of occurrence of gray contour.

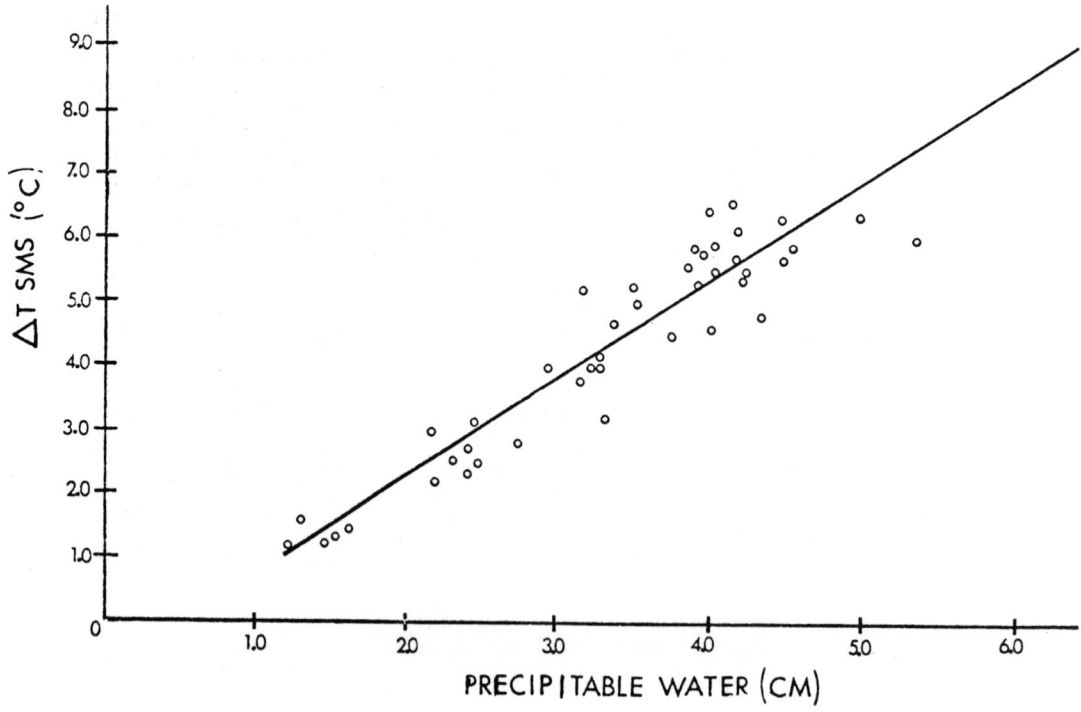

Fig 4. SMS ΔT as a function of precipitable water.

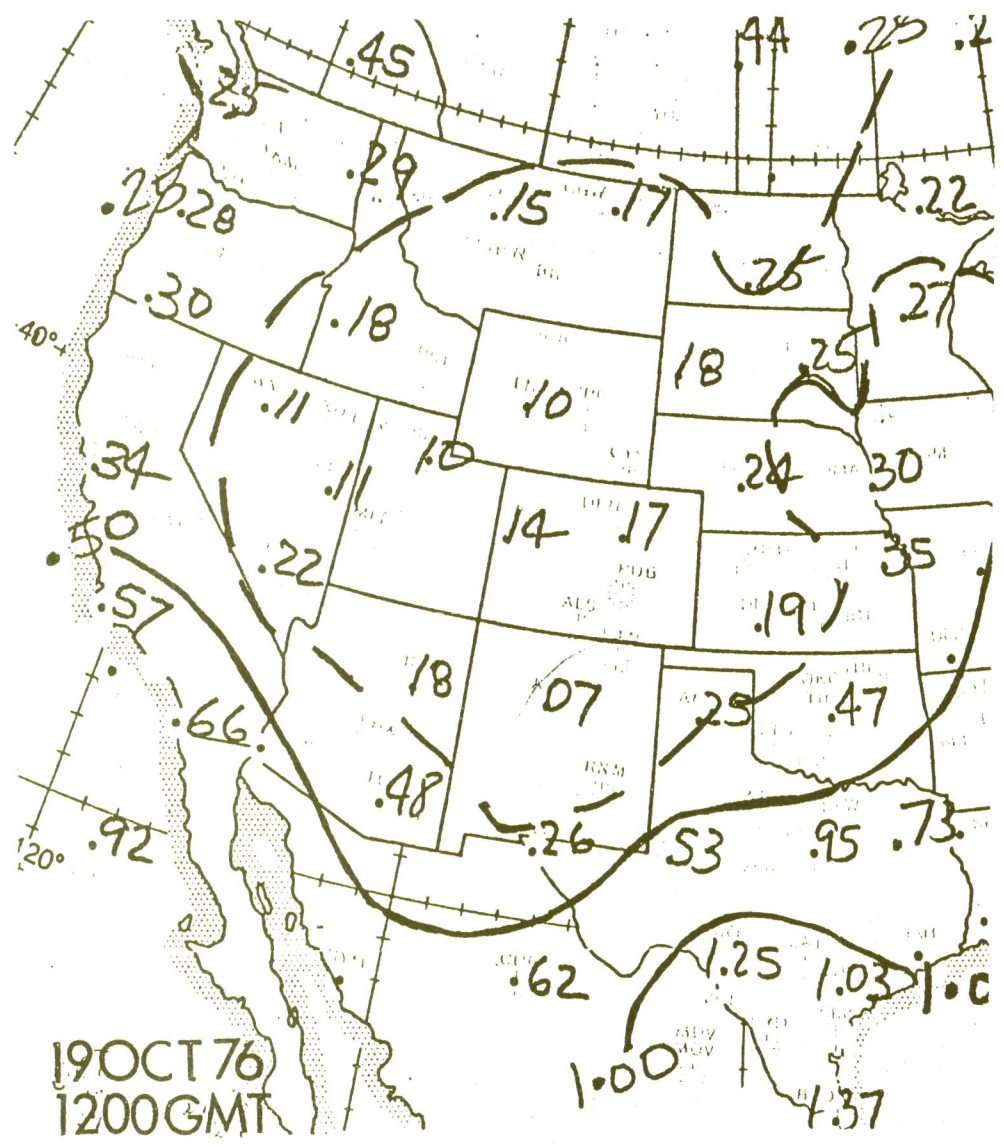

Fig 5. Precipitable water chart - 19 Oct 76, 1200 GMT.

U.S. DEPARTMENT OF COMMERCE

National Weather Service/National Environmental Satellite Service
SATELLITE APPLICATIONS INFORMATION NOTE 76/7*

OCEAN THERMAL FEATURES AS SEEN FROM GOES-1

Stephen R. Baig
Satellite Field Services Station, Miami, FL

The National Weather Service has recently introduced a "Gulf Stream Wall Bulletin" as an adjunct to regular East Coast weather broadcasts for mariners. The bulletin provides the location of the western edge (cyclonic edge) of the Gulf Stream.

The National Environmental Satellite Service makes input to this product with thermal infrared data from both the NOAA-4 (polar orbiting) and GOES-1 (geostationary) satellites. This note will describe some of the Miami SFSS experience with GOES-1 from the marine viewpoint.

There can be little doubt that for non-spatially limiting features, imagery from a geostationary satellite is much preferable to that from a polar orbiting satellite, at least where there are no features to provide easy registration of latitude and longitude.

The second and perhaps most important advantage of the GOES-1 data is that it provides an image every half-hour instead of every twelve hours of the entire Gulf Stream region. Ocean features do not move with speeds which would require images more often than once every few days, were it not for the overlying clouds. When a few hours of GOES-1 imagery are used to prepare an animated movie the data seem to come alive! Small cloud-free areas move along with the air flow, revealing the positions of thermal features in the nearly stationary water below. The satellite analyst's eye persistence integrates this information, allowing the thermal features in the water to be described with a continuous contour.

The images used for the preparation of the animated movies are "enhanced" to better display the weak thermal gradients found in the ocean. NESS oceanographers have prepared a number of different enhancements for various parts of the ocean. For the Gulf Stream and adjacent waters the most useful enhancement currently available is labeled "S" (The type of enhancement displayed in a GOES image is indicated in the picture header.) A description of the S-curve enhancement is included in the "Enhancement Kit" available at all WSFO's.

It is expected that, as more interest is generated within the marine community, the marine forecasts will be asked to supply not only the Gulf Stream Wall Bulletin but also more detailed information about

---
* We've been asked by subscribers to number the notes serially for each calendar year. So far in 1976 we've distributed 2 Information Notes in each of the first 3 months and this one in April for a total of 7.

thermal features in smaller areas, and the changes features undergo after the Bulletin has been issued. Indeed, the increased number of observations provided by GOES-1 imagery already has shown us a higher variability in the Gulf Stream position than had been generally recognized.

Occasionally the cloud conspires to clearly reveal large portions of the oceans. An example is shown in Figure 1. The Gulf Stream system is well defined from near the Georgia coastal zone downstream to a position south of Nova Scotia. A series of small edge waves can be seen along the cyclonic edge.

Southeast of Cape Hatteras a pool of cool water is nearly surrounded by a warm band. Two well defined warm eddies can easily be seen in the slope water between the cyclonic edge and Nova Scotia. Data from recent research cruises into similar eddies shows that the water is uniformly warm right to the bottom of such eddies, (Apel & Byrne). Many such features can be tracked from day-to-day.

In the Gulf of Mexico a good portion of the Eastern Gulf Loop Current is visible. This water bulges into the Gulf and then breaks off, forming a gigantic warm eddy. The process repeats on an annual cycle (Maul, 1975). Knowledge of the extent and position of the features described above are of immense value to merchant shipping as well as fisheries interests.

In the Eastern Pacific two patches of cold water can be seen extending out from the shore. One, near 16N98W, is the result of the classic Tehuantepecer, cold air being funneled through the mountain pass of the Isthmus of Tehuantepec. The second patch extends seaward from near 12N86W. In this case air is funneled over the Lake Managua region. These low level jets cause a seaward displacement of the usually warm surface water and brings colder deep water to the surface. In both areas the marine forecaster can therefore infer the direction and relative magnitude of the low level winds. Note also that the wind stress curls the cold water to the right.

It should be noted that only a few hours before this GOES image was made the entire Gulf Stream system was covered by cloud. A few hours after this iamge was made, cloud again obscured the Gulf Stream. By monitoring the images on a 30-minute schedule the marine forecaster can request ocean enhanced images timed to provide him with a maximum of information. By contacting the oceanographer or hydrologist at the NESS/SFSS serving him the marine forecaster can obtain other information which will be useful in his forecast speciality.

Figure 2 shows the analysis made from the movie loop prepared using a series of images including figure 1. These loops are analyzed every day at Miami. On Tuesdays a final weekly analysis is prepared, on tracing paper, using Tuesday's data. If a portion of the area is obscured, then data from Monday is used, and so on. In this way the latest available data is included in the final weekly analysis. If an area is obscured for more than one week, no data is shown on the analysis. By hanging the most recent analysis over those of the previous weeks, changes can easily be seen and data from previous weeks used to estimate positions in the breaks in the most recent analysis.

REFERENCES

Apel, J. and Byrne, M., Personal Communication.

Maul, G. A., "An Evaluation of the Use of the Earth Resources Technology Satellite for Observing Ocean Current Boundaries in the Gulf Stream System', NOAA Tech Report, ERL 335-AOML 18, 1975.

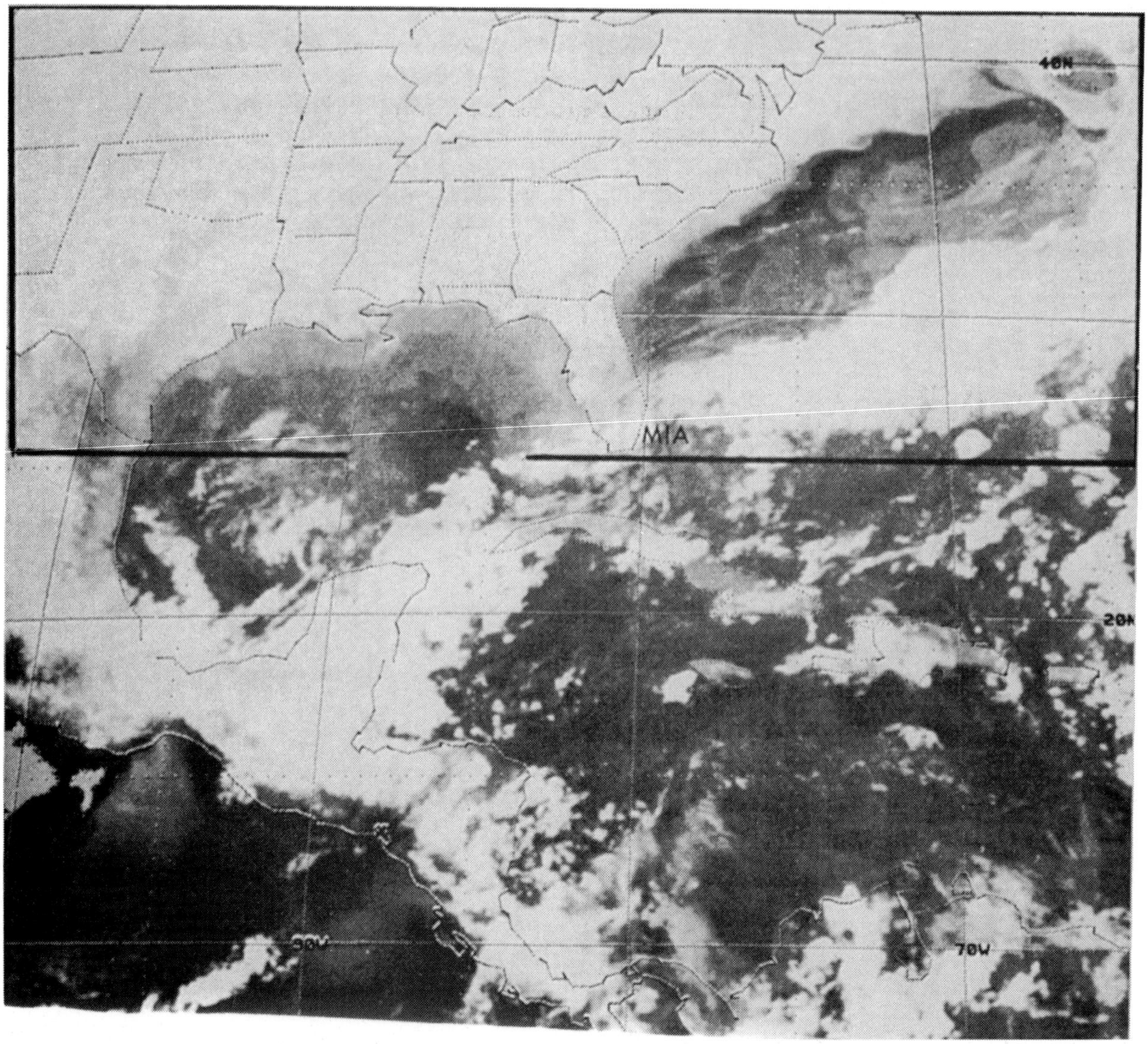

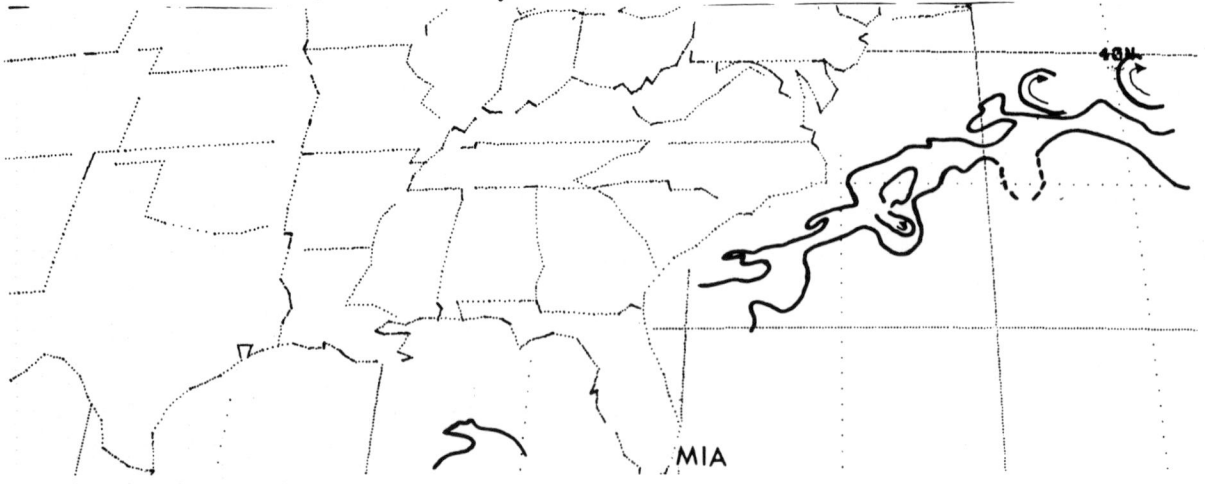

Figure 1. GOES-1 Enhanced IR (S-Curve) showing oceanographic features at 0300 GMT, 28 February 1976.

Figure 2. Gulf Stream and Loop Current analysis derived from data in Figure 1.

CHAPTER 2

SYNOPTIC ANALYSIS

This chapter relates satellite imagery to large-scale synoptic and dynamic weather processes and systems. Significant meteorological processes and features show up as discontinuities, or characteristic shapes on imagery. These meteorological features usually behave and evolve according to specific patterns, and according to specific laws of fluid-mechanics. Expected evolution, deformation, and movement of systems must be identified, extrapolated, and compared to expected positions on numerical guidance before a useful forecast can be made.

References in this chapter relate features on satellite imagery to basic atmospheric behavior. They should also provide a framework for identifying, explaining, and understanding large-scale atmospheric features affecting a forecast area. The first four papers discuss synoptic scale cloud patterns and relate these patterns to upper level wind flow. The remaining four papers discuss cyclogenesis.

> 2-A Synoptic Scale Cloud Systems. Roger Weldon, NESDIS Training Notes, Satellite Applications Laboratory, National Oceanic and Atmospheric Administration, U.S. Department of Commerce, Washington, D.C., September 1983, 35pp.

This reference provides a basic statement of meteorological dynamics and processes from the perspective of satellite imagery interpretation. Fundamental principles, as seen on imagery, are described qualitatively in six sections. The first section describes movements and processes with respect to the atmosphere, rather than to the earth's surface. The second section relates features on imagery to these atmospheric motions. The third and fourth sections identify and explain the fundamental large-scale cloud patterns on imagery. The fifth section explains both baroclinic and large-scale convective energy storage and transfer, their interrelationship, and their influence on storm evolution. Finally, upper level cut-off systems and the development and evolution of atmospheric blocking patterns are explained.

In this paper, the Norwegian concept of fronts, waves, and air masses has been modified by the concept of baroclinic energy storage and transferral. For example, a surface front can be thought of as the low-level portion of a deformation zone, traveling along with the windflow. In addition, a segment of a jet stream can be thought of as kinetic energy that results from processes that are storing or have stored baroclinic potential energy upwind.

SYNOPTIC ANALYSIS

    2-B   Part IV Cloud Patterns and the Upper Air Wind Field. Roger Weldon, NESDIS Training Notes, Satellite Applications Laboratory, National Oceanic and Atmospheric Administration, U.S. Department of Commerce, Washington, D.C., March 1979, 3-13.

This reference supplements the previous one by describing jet streams and vorticity lobes. Different types of jet streams are first explained dynamically. Finally, four rules are provided for locating jet streams on satellite imagery. This, and the previous reference should provide a basic understanding of cloud dynamics as seen from the satellite perspective.

    2-C   Airflow Through Mid-Latitude Cyclones and the Comma Cloud Pattern. Toby N. Carlson, Monthly Weather Review, 108 (10), October 1980, 1498-1509.

This reference presents another way of looking at meteorological processes and relating them to satellite imagery. Toby Carlson looks upon three different air masses, that interact in most mid-latitude cyclones, as conveyor belts that flow over and move under one another. The system evolves and travels as a consequence of these movements. Vertical and horizontal temperature, moisture, and precipitation distributions are explained.

It must be stressed that the three ways of looking at weather systems are complementary. If time is taken to carefully visualize each, the reader will develop a deep understanding of how the atmosphere tends to behave and how features should appear and evolve on numerical guidance and on imagery.

    2-D   Satellite Interpretation. Eugene M. Weber and Steven Wilderotter, Third Weather Wing/Technical Note-81/001, Aerospace Sciences Division, Offutt AFB, NE, December 1981, 28-35.

This reference describes the identification of standard atmospheric features, and follows their evolution as depicted in the imagery. This particular reference illustrates the location of upper ridges and troughs on imagery. Guidelines are provided for estimating height contours, using imagery. Numerous examples are provided.

    2-E   Surface Cyclogenesis as Indicated by Satellite Imagery. Frank J. Smigielski and Gary P. Ellrod, NOAA Technical Memorandum NESDIS 9, National Oceanic and Atmospheric Administration, U.S. Department of Commerce, Washington, D.C., March 1985, 30pp.

SYNOPTIC ANALYSIS

This reference describes and illustrates cyclogenesis as depicted on imagery. Indications and types of cyclogenesis, as well as rates of intensification are discussed. Qualitative evaluation follows, including estimating surface pressures, and locating and estimating the strongest surface winds.

2-F First Glances can be Misleading in Locating Vorticity Centers. Carl E. Weiss, Satellite Applications Information Note 2/77, NWS/NESS, U.S. Department of Commerce, Washington, D.C., 1977, 5pp.

This reference provides guidelines for locating vorticity centers on imagery. Common problems in locating these circulation centers are mentioned. The most prevalent problem involves detecting and tracking vorticity centers that are commonly located in cloud-free areas.

2-G An Oceanic Cyclogenesis - Its Cloud Pattern Interpretation. Roger B. Weldon, Satellite Applications Information Note 77/7, NWS/NESS, U.S. Department of Commerce, Washington, D.C., 1977, 11pp.

This reference provides a well articulated explanation of energy transfer in the atmosphere. It also provides excellent illustrated examples of cyclogenesis and subsequent weakening.

2-H Extratropical Cyclogenesis over the Gulf of Mexico. Brian E. Heckman and A. H. Thompson, Proceedings of the Conference on Weather Forecasting and Analysis and Aviation Meteorology, October 16-19, 1978, Silver Spring, MD., Americal Meteorological Society, Boston, MA, 118-124.

This paper summarizes eight cases of early cyclogenesis over a data-sparse area. The first indications of cyclogenesis appear as systematic cloud pattern changes on satellite imagery 18 hours before a closed 2 mb isobar appears on surface charts. These characteristic changes, and indications of continued cyclogenesis are described and illustrated.

SYNOPTIC SCALE CLOUD SYSTEMS

Roger Weldon

September 1983

The following notes illustrate cloud patterns and features observed on satellite imagery, and their relationship to the wind and density fields. The relationships shown by drawings and briefly discussed are primarily empirically derived. The figures used were selected from unpublished training notes written by the author; but, the brief text and explanations of figures were rewritten. These notes represent a brief summary of the subject material, and are intended to support training sessions using imagery in slide and film loop form.

Contents

PART ONE	BASIC ASPECTS OF THE WIND FIELD WHICH AFFECT CLOUD PATTERNS AND THEIR EVOLUTION	1
PART TWO	EXAMPLES OF DEFORMATION ZONE CLOUD FEATURES	7
PART THREE	THE COMMA PATTERN	11
PART FOUR	THE "BAROCLINIC LEAF" CLOUD PATTERN	21
PART FIVE	THE D.A.V.E. MODEL AND CLOUD PATTERN EVOLUTION	28
PART SIX	CUT-OFF LOW CIRCULATIONS AND CLOUD PATTERNS	32

PART ONE BASIC ASPECTS OF THE WIND FIELD WHICH AFFECT CLOUD PATTERNS AND THEIR EVOLUTION

Our extensive studies of cloud systems using satellite imagery in time lapse motion have revealed certain aspects of the wind field which are very significant to understanding cloud patterns and features. These are:

   a. Rotation and vorticity
   b. Deformation and "deformation zones"
   c. Speed Maxima

The three aspects and the fundamental relationships among them are illustrated in the following pages. In each case, it is the "relative motion" of part of the air with respect to other parts that is related best to cloud patterns, not the motion with respect to the earth which is explicitly shown by the "wind" field.

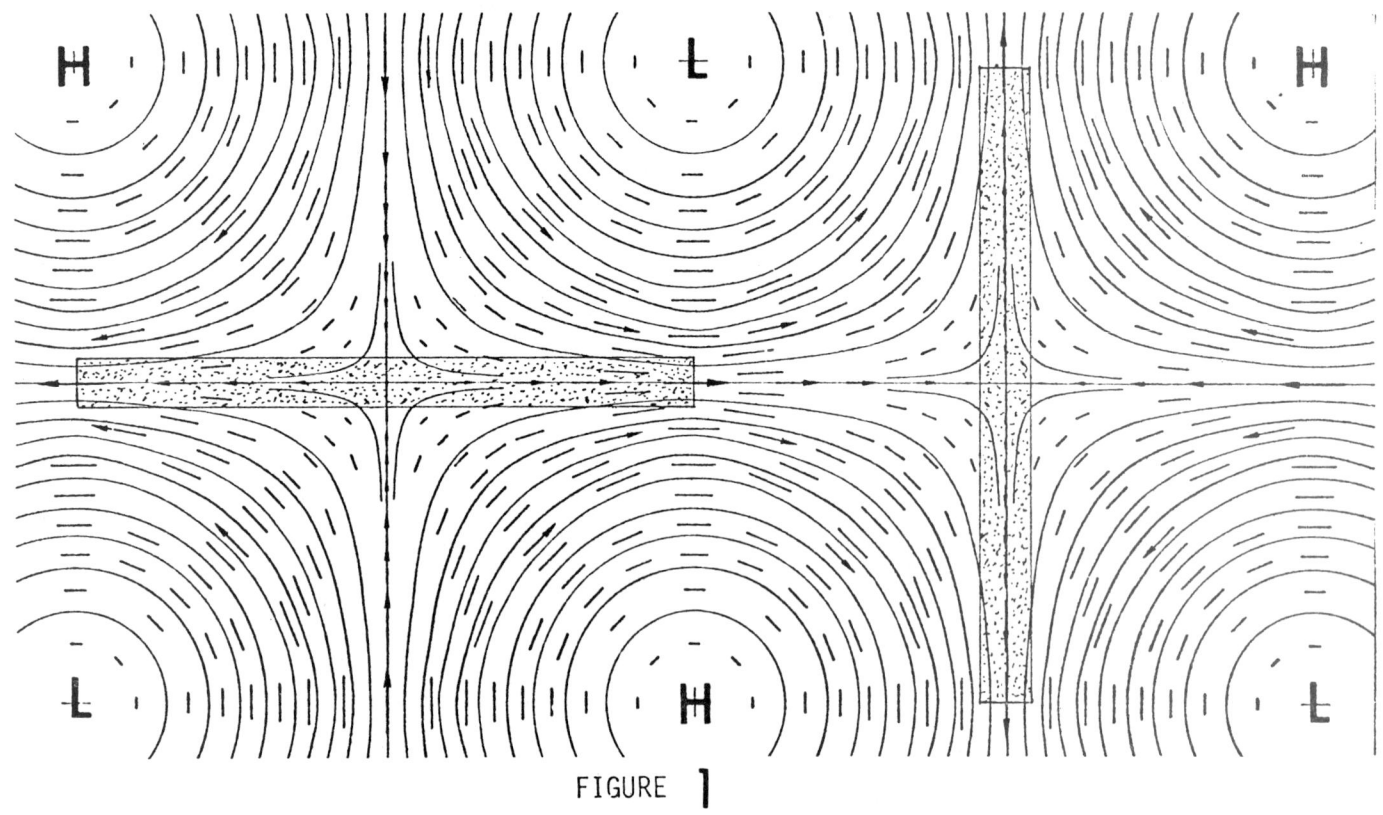

FIGURE 1

FIGURE 1

The drawing shows a field of motion. The long thin lines are streamlines. The short straight lines are motion vectors. The speed is proportional to the length of the vector. Shown are two hyperbolic regions or cols surrounded by parts of three cyclonic and three anticyclonic circulation systems.

The two elongated shaded regions located along the stretching axes of the cols are defined as "DEFORMATION ZONES". A large number of cloud and moisture features observed on satellite imagery, such as lines, bands, and boundaries, are located parallel to deformation zones in the associated motion field.

If the circulation systems shown in Figure 1 represent air motions, and are stationary with respect to the earth; the vectors and streamlines represent the WIND field.

FIGURE 2

This shows the same circulation systems as in Figure 1, but they are all moving or translating at the same speed toward the right side. The long thin lines are the same as the streamlines of Figure 1; but, in this case, they represent the "RELATIVE MOTION" of the air with respect to the rest of the air.

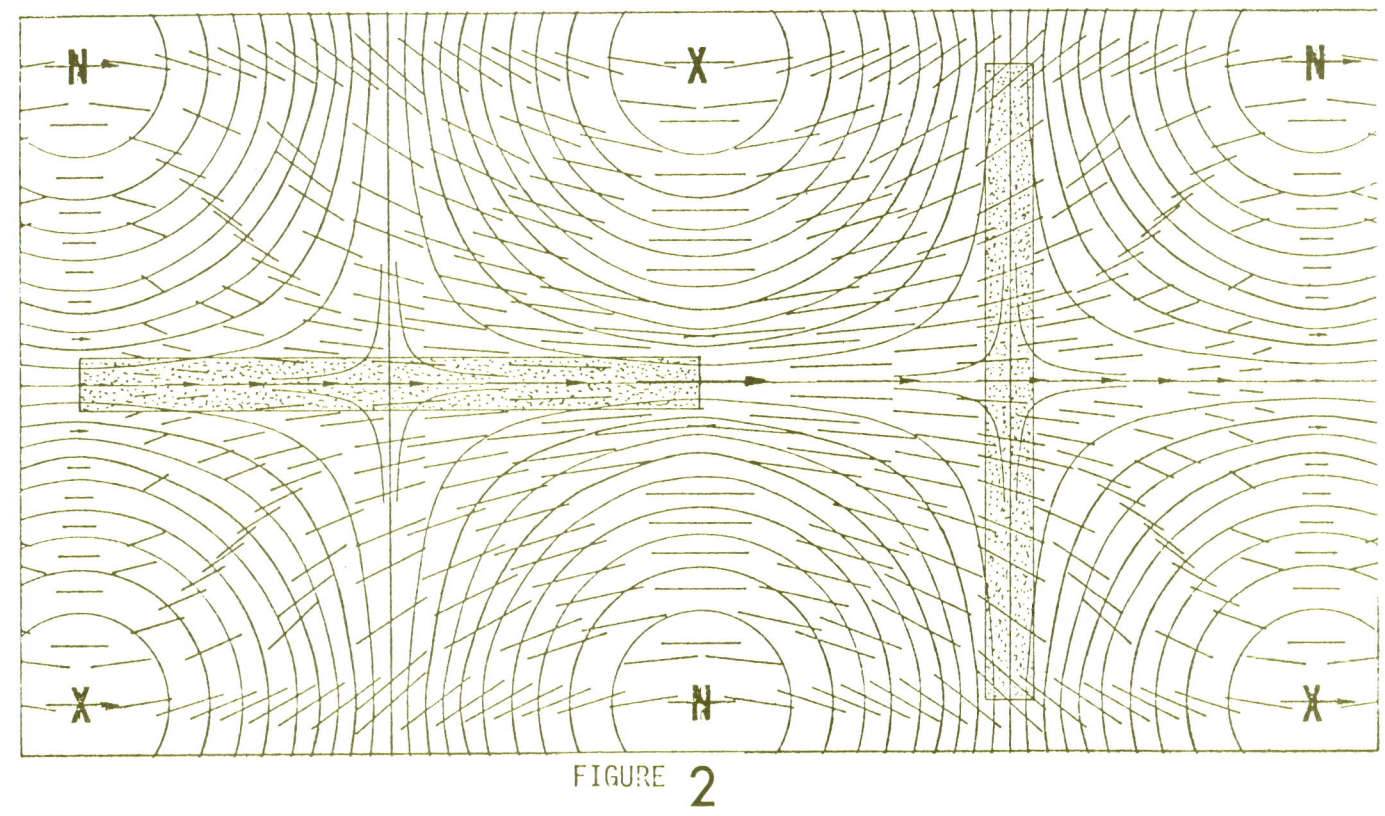

FIGURE 2

The short lines, or wind vectors, must in this case include the velocity of motion toward the right. When the systems are moving, the wind vectors do not explicitly show the circulation centers or hyperbolic zones as in Figure 1. However, the circulations are still present in the relative motion field. Cloud patterns would be related to the relative motions. They would not be as easily related to the wind vectors, or streamlines drawn to them. If clouds or moisture were present within the wind field of Figure 2, shown by the vectors, significant bands or boundaries would likely along the deformation zones, and circulation centers would be revealed at the locations marked "X" and "N". These locations represent centers of cyclonic and anticyclonic vorticity, instead of cyclone and anticyclone centers as in Figure 1.

The central vector of Figure 2 is the maximum speed. The deformation zones form a "T" pattern with a definite fundamental relationship to the speed maximum. The "trailing" deformation zone is parallel and to the rear of the speed maximum, forming the stem of the "T". The "leading" deformation zone is orthogonal and in advance of the speed maximum, forming the cross bar of the "T". Maximum cyclonic vorticity is to the left of the speed maximum (Northern Hemisphere), and maximum anticyclonic vorticity is to the right side. This relationship among the rotation, deformation zones, and the speed maximum is a very basic one to fluid motion. It is important to understanding satellite imagery at all scales and all latitudes.

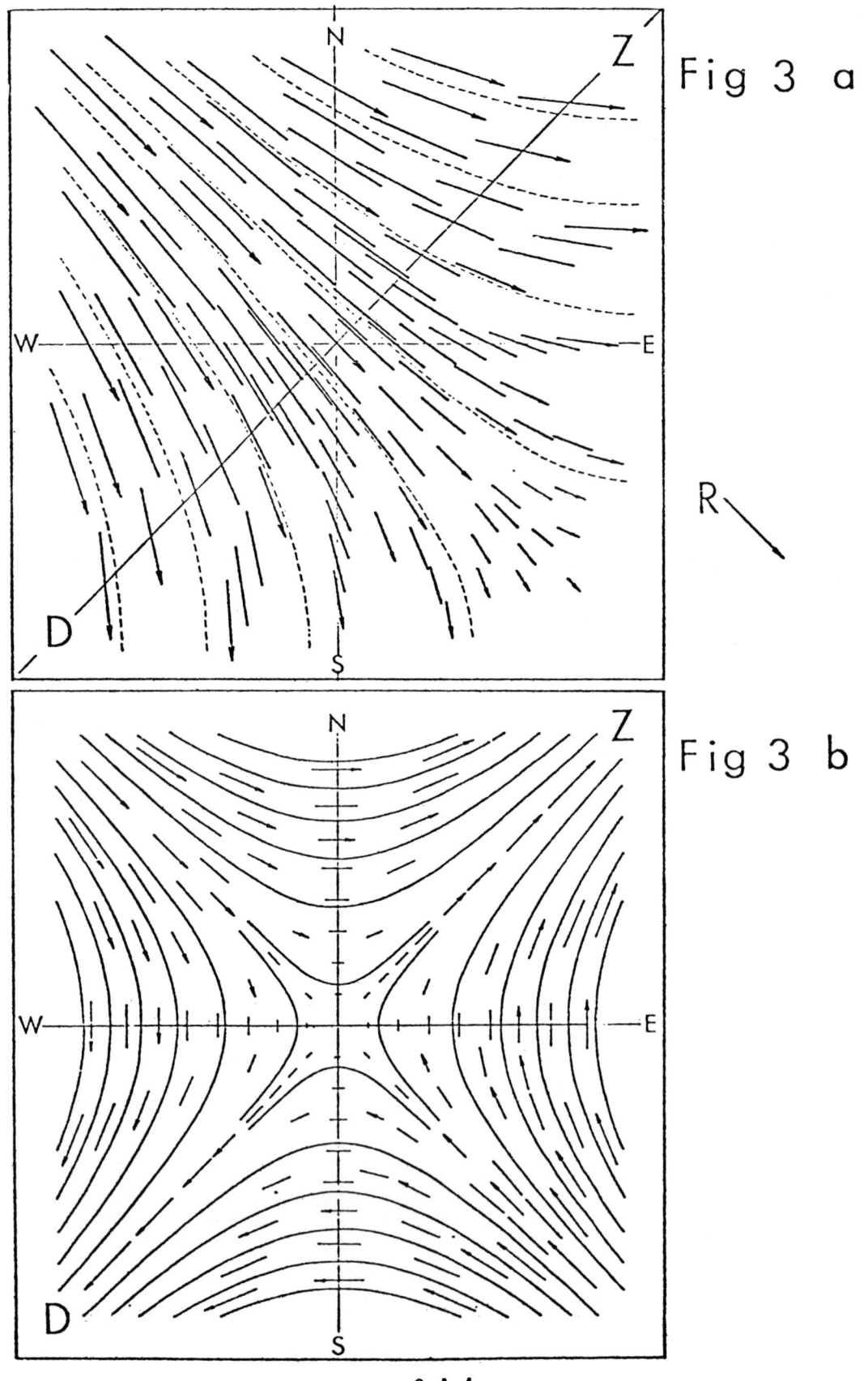

Fig 3 a

Fig 3 b

FIGURE 3

Figure 3a shows a field of northwesterly winds. The maximum wind speed is in the northwest corner of the drawing. Downstream from that point, the flow is difluent and speeds decrease. If clouds, moisture, dust, or some other substance were introduced into the flow at the northwest corner, how would it behave with respect to the clear air in advance? Perhaps it would become more and more diffuse as it spreads southeastward. Many such cases viewed in time lapse satellite imagery have indicated otherwise. Both the upstream air containing the substance, and the downstream clear air spreads to the sides as it moves downstream. But, the downstream clear air tends to fall back relative to the upstream air which is moving more rapidly.

To examine the relative motion field involved, the mean resultant wind vector of all those in Figure 3a is determined. It is shown as vector "R" to the right. If this mean vector is subtracted from all those of Figure 3a, the motion field in Figure 3b is obtained. This relative motion field reveals a hyperbolic zone with its stretching axis, or deformation zone, along line "D-Z", This indicates that the upstream air, while overtaking the downstream air, is itself becoming slower and will approach the downstream air asymptotically at axis "D-Z". A distinct boundary is formed by the deformation process at "D-Z". The boundary will be moving toward the southeast with the speed of vector "R".

Figure 3b shows the relative motion of the air with respect to the other air. This motion "system" is moving with vector "R". Figure 3a shows the WIND field, or the motion of each piece of air with respect to the earth's surface. Patterns seen on the satellite imagery evolve according the to relative motions. For that reason, they are not always easily related to the associated wind field.

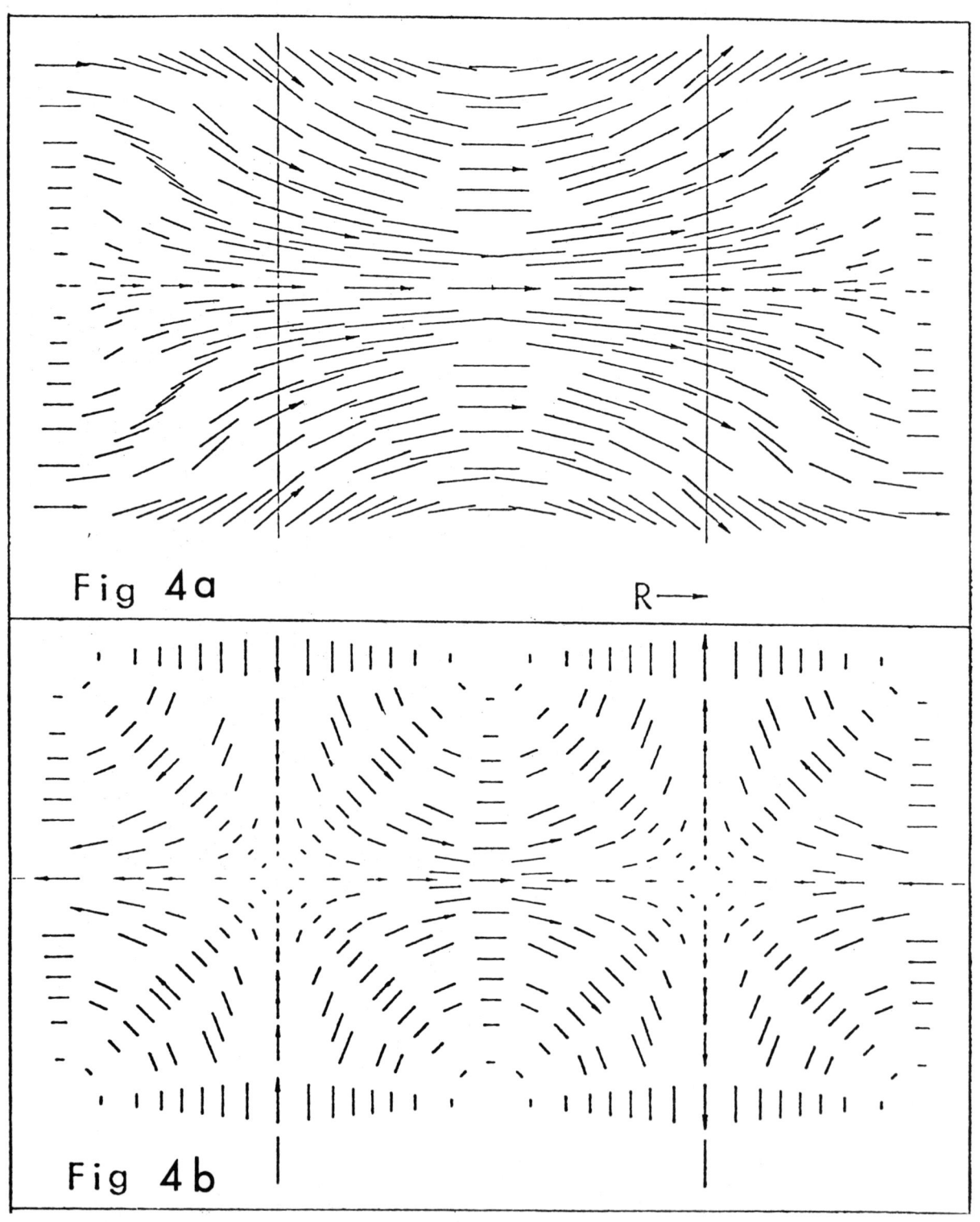

Figure 4

This illustrates the same concept as in Figure 3, but with a more complete circulation system involved. Winds are shown in Figure 4a. If the mean wind - vector "R" - is subtracted from the vectors in 4a, the relative motion field in Figure 4b results. The relative motion field reveals circulation centers and deformation zones forming the "T" pattern with respect to the speed maximum at the center of Figure 4a.

PART TWO   EXAMPLES OF DEFORMATION ZONE CLOUD FEATURES

FIGURE 5

CLOUD BANDS along Deformation Zones. The solid lines are streamlines of the wind field. The band of clouds in Figure 5a has a hyperbolic zone in the winds. The feature would not be moving rapidly. The band in Figure 5b is moving toward the southeast, and the one in Figure 5c toward the north.

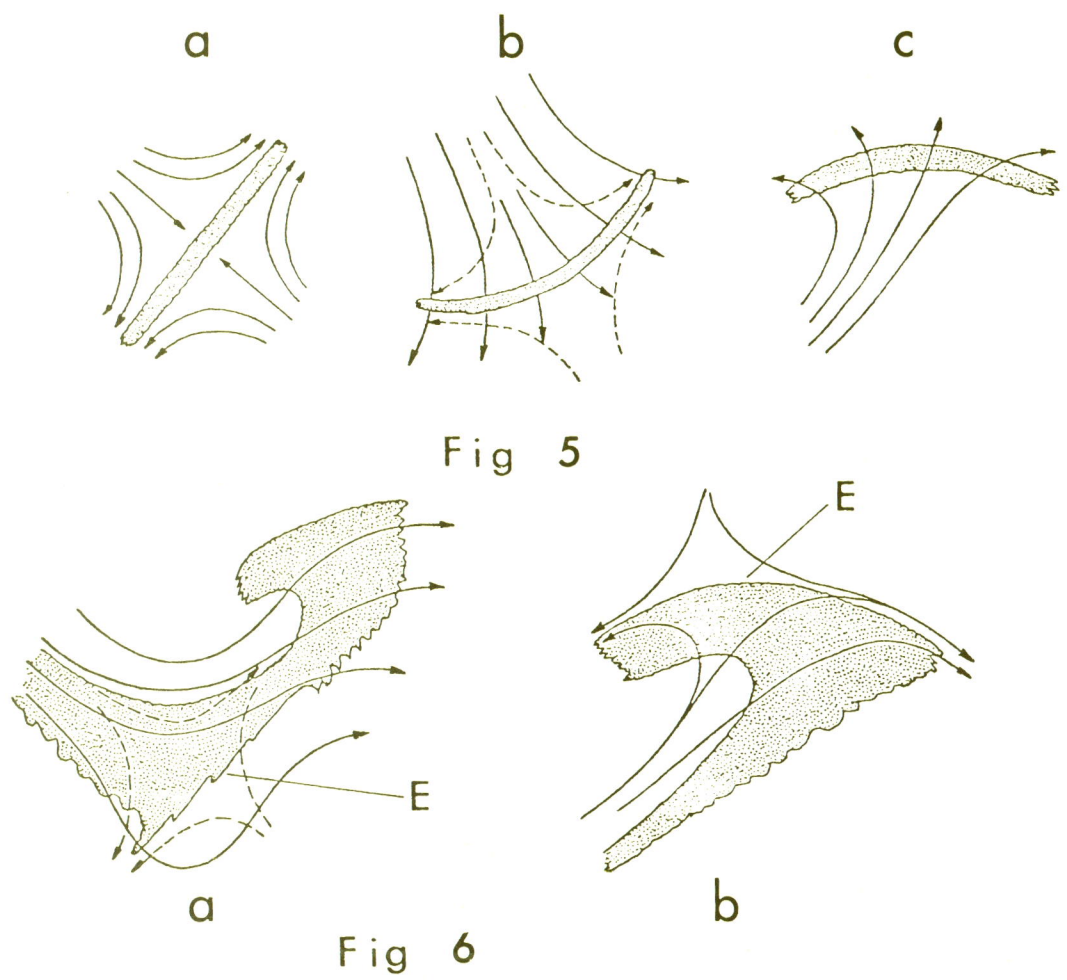

Fig 5

Fig 6

FIGURE 6

CLOUD BOUNDARIES along Deformation Zones. The cloud patterns shown are those of synoptic scale midlatitude storms. Two common locations for deformation zone borders are shown. In Figure 6a, the deformation zone boundary is at "E". It is common for this boundary to be moving. In that case, the southern trough is not likely a "closed" circulation. The hyperbolic motions are in the relative flow, and would be defined in the vorticity field. The boundary at "E" in Figure 6b is more common, and is also frequently associated with a col or hyperbolic zone in the upper level winds.

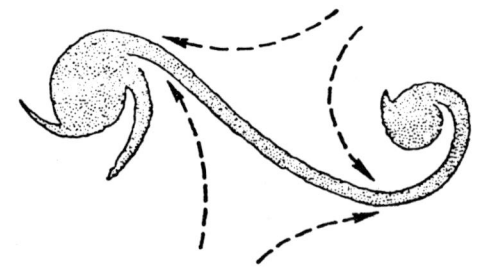

Fig 7a

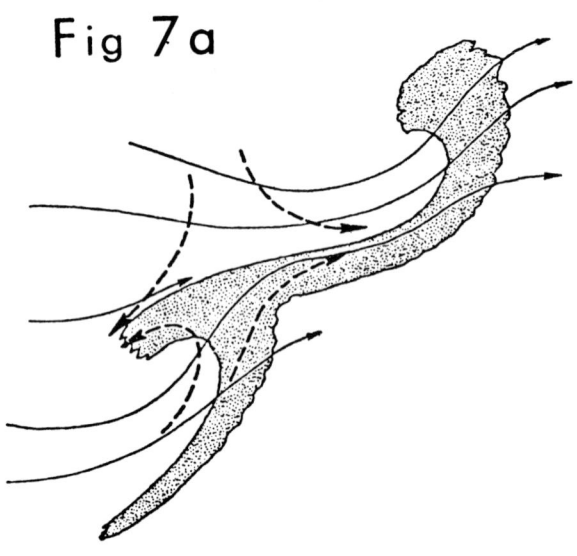

Fig 7b

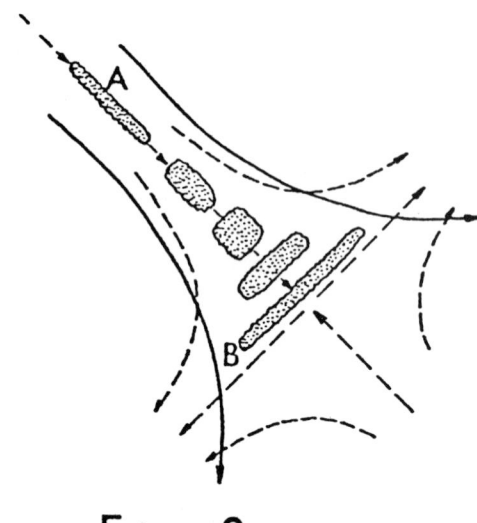

Fig 8

FIGURE 7

Deformation Zone CONNECTING BANDS. Figure 7a shows a "connecting band" through the deformation zone between two tropical cyclones. A hyperbolic zone in the wind field is likely. The connecting band in Figure 7b is between two short wave storm systems in the mid latitude westerlies. A hyperbolic zone in the associated wind field is less likely since the systems are usually moving rapidly.

Figure 8

Deformation CLOUD PATTERN. When viewed in time-lapse motion of satellite images, cloud "A" may change shape and move into the position of cloud "B". As that happens, cloud "B" has dissipated, and new clouds have formed upstream. An array of different clouds in different phases of deformation can be seen in a single image.

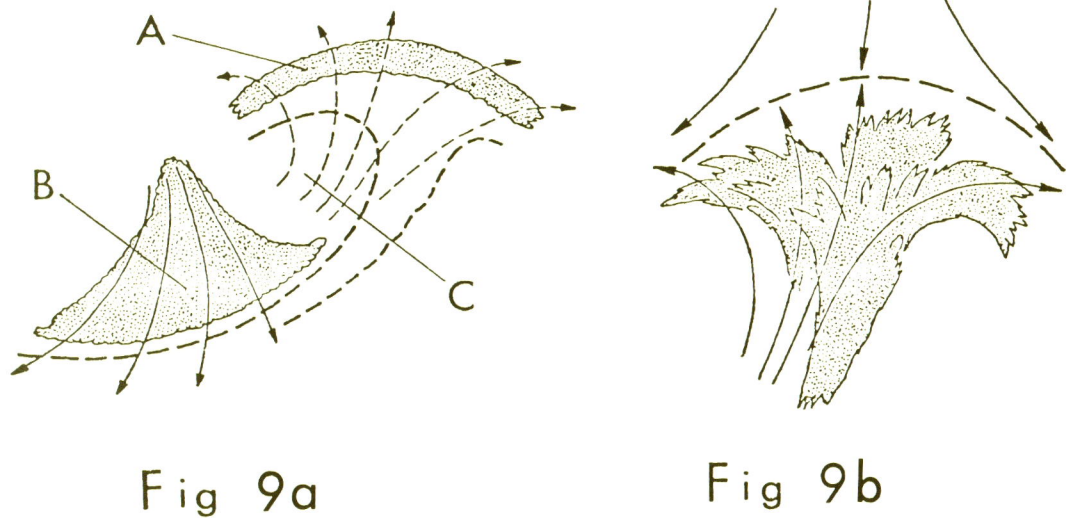

Figure 9

The drawings in Figure 9 show deformation zone cloud features that are often seen with mid tropospheric cold lows or "cut-off" lows. In Figure 9a, the deformation band at "A" is located in the southerly winds of the upper level flow. The deformation boundary at "B" is in the northerly winds of the low level flow. The two cloud features define the location of the cold low. If deep convection such as thunderstorms are occurring under the upper level cyclonic circulation, a pattern in the cloud debris, such as that shown in Figure 9b, will often be seen. Sometimes a well defined boundary will form along the dashed line at the deformation zone.

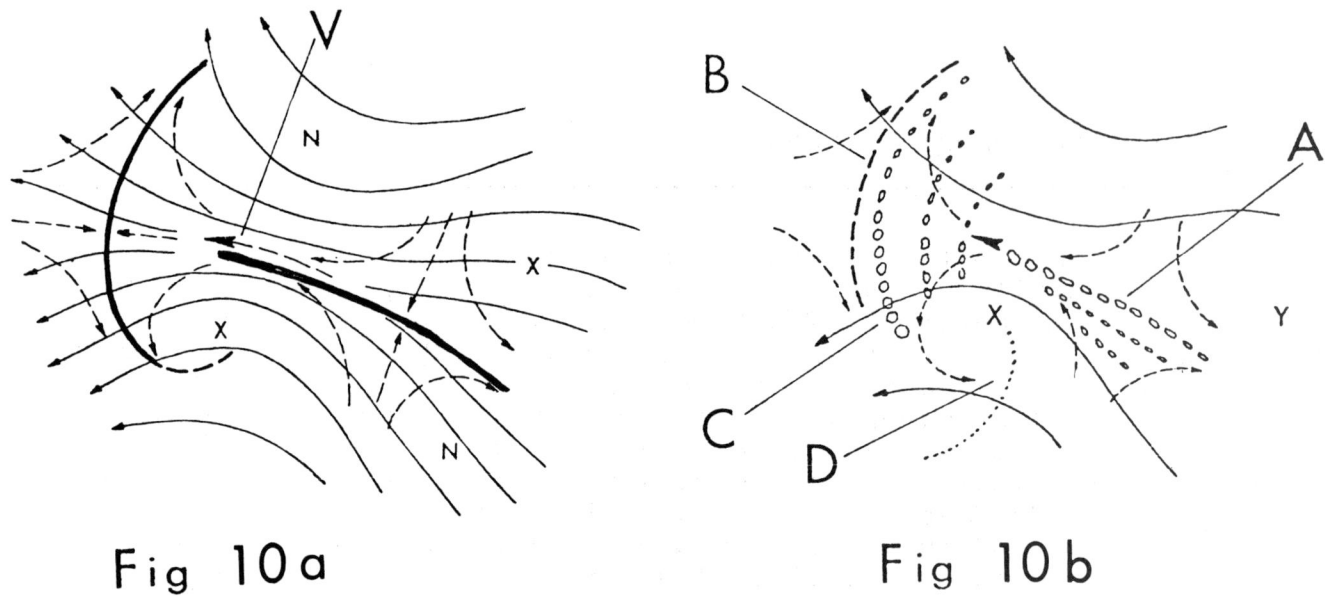

Fig 10a   Fig 10b

The drawings in Figure 10 illustrate cloud systems frequently observed in the low level easterlies at subtropical and tropical latitudes. The significant features are two deformation zones and a speed maximum in the easterly flow. The two thick lines in Figure 10a represent cloud bands. Thin solid lines are streamlines, and dashed lines with arrows indicate relative motions in the flow.

The western band in Figure 10a is associated with the "LEADING Deformation Zone" in advance of the speed maximum at "V", and the eastern cloud band is with the "TRAILING Deformation Zone" to the rear and parallel to the speed maximum. With this Northern Hemisphere depiction, cyclonic rotation is to the left of the speed maximum on south side; and anticyclonic rotation is to the right.

The deformation zones with such systems often also act as boundaries between different kinds of low level cloud conditions. Often an elongated band of very stable conditions is found along the "leading" deformation zone, with unstable conditions on its east side in advance of the speed maximum. Lines of convection in this region are frequently oriented parallel to the "leading" deformation zone, as indicated in Figure 10b. The "leading" band is often better defined in the clouds along its southern end to the left front quadrant of the speed maximum. If deep convective activity forms and persists along the leading band, it is most likely along that portion. At times, when the system becomes deformed by shears from the environmental circulation systems, the southern half of the "leading" deformation zone along with the "trailing" deformation zone forms an inverted "V" shape.

If the "trailing" deformation zone acts as a boundary between cloud types, the area south of the band is often clear or with very small cumulus; and north of the band, is of stratocumulus with debris spreading in the marine inversion.

If the easterlies are deep, and the upper level flow is consistent with the lower level perturbations, deep convection may form and persist between the vorticity maximum at "X" and the speed maximum. In that case, a low level warm core low development is likely.

PART THREE   THE COMMA PATTERN

Perhaps the most frequently applied term describing cloud patterns on satellite imagery has been the "comma", "comma cloud", or "cloud comma" pattern. The term - derived from the shape of the punctuation mark - was first applied to small scale convective cloud systems often found under the cold air aloft to the left of the jet stream over oceans. These were not normally detected in the sparse conventional data, and did not have a name. The cloud shape was very similar to the punctuation comma with the "S" shape rear border, wide "head" and narrow "tail" feature.

With time, and the increased understanding of the related wind field and pattern evolution involved, the term was used to describe a much larger variety of cloud systems, from large winter storms to small convective systems.

The primary feature of the comma pattern is the normally well defined "S" shaped upstream border. The evolution into this shape is best related to differential rotation in the relative motion field in which the border is propagating.

Cloud comma systems are illustrated in the following figures.

FIGURE 11

FIGURE 11

The drawings in Figure 11 represent a time sequence of change from the top left to the lower right. The first drawing shows a circular area with a straight border dividing the area into halves. If the circular area is allowed to rotate differentially, so that the maximum tangential velocity is along the inner circle; the border will become deformed into an "S" shape. This is the primary process involved in the formation of comma cloud systems. The flow is not parallel to the border, but across the border a varying speeds, so that one part moves at a different speed than the part adjacent to it. Note that the portion of the border inside the maximum speed circle rotates cyclonically with time, and the portion outside that circle rotates anticyclonically. The amount of rotation varies along the border with the maximum cyclonic rotation at the inflection point.

If the rotational motion illustrated in Figure 11 represented a wind field, the area inside the circle would be a "closed" low or cyclone with varying vorticity. The center of the low and the maximum cyclonic vorticity would be at the center of the circles. If we consider only the first four drawings, and assume that the system were translating rapidly eastward while undergoing the differential rotation; then, the part of the border that deforms westward may not actually be moving that direction, but only in a

relative sense with respect to the other parts of the border.
It may be falling back with respect to the other points, but
still moving eastward. In that case, there would be no "closed"
circulation - no east winds - but a vorticity center or point of
maximum cyclonic vorticity would be located at the inflection
point of the border.

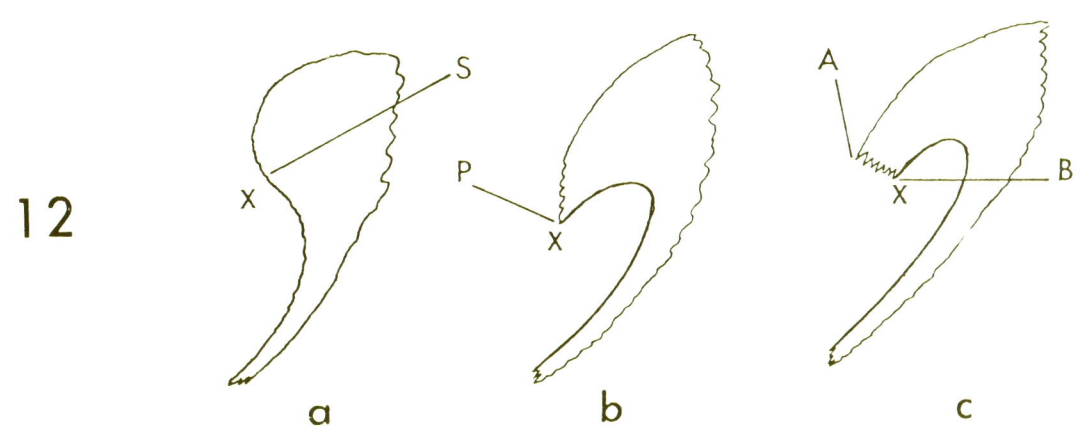

FIGURE 12

The three drawings show typical cloud comma patterns observed
in satellite imagery. The pattern in Figure 12a has an "S" shaped
upstream border similar to the punctuation mark comma. More common
is the pattern in Figure 12b, in which the inflection point is
replaced by a single apex or sharp corner on the cloud border.
Even more frequently observed, is the pattern in Figure 12c, with
a double point or double corner separating the well defined convex
and concave portions of the border. If the cloud systems shown are
in the westerlies and of middle or high cloud tops, the "X" on
the drawings indicates the most likely location of the maximum
cyclonic vorticity at 500mb. The relationships were determined
by observations of satellite imagery and 500mb vorticity analyses.

13 a

13 b

14 a

14 b

14 c

15 a

15 b

15 c

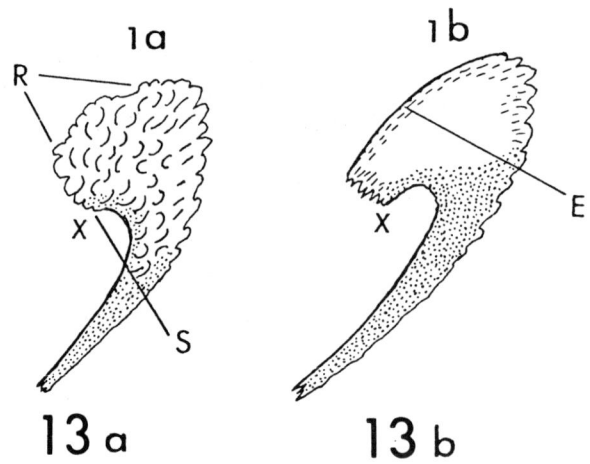

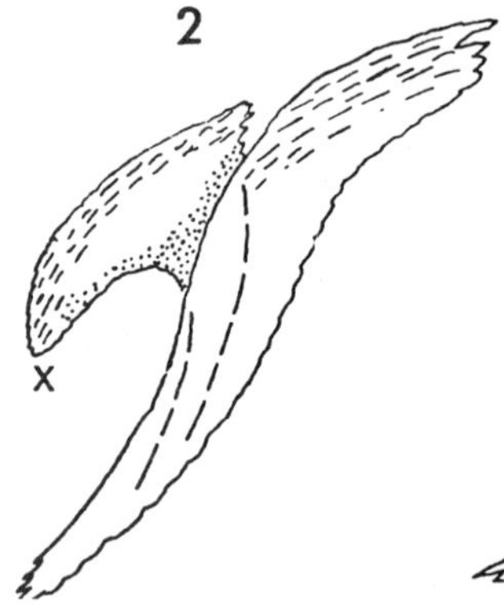

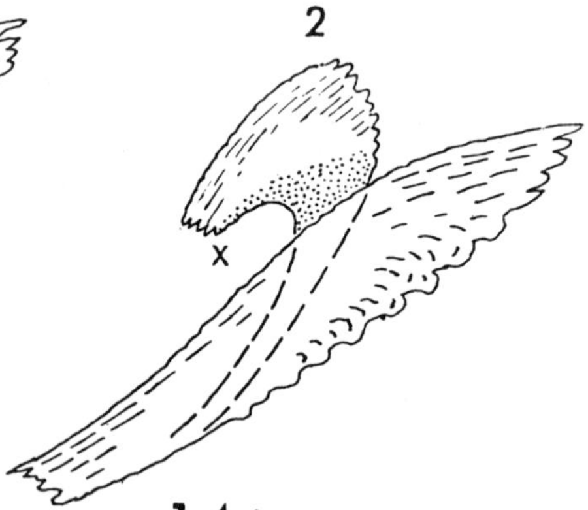

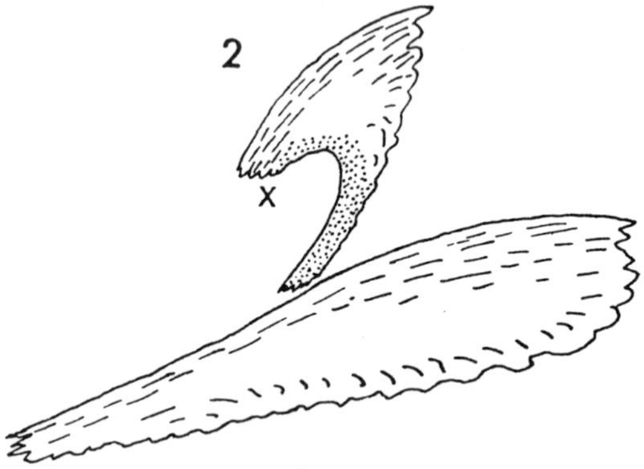

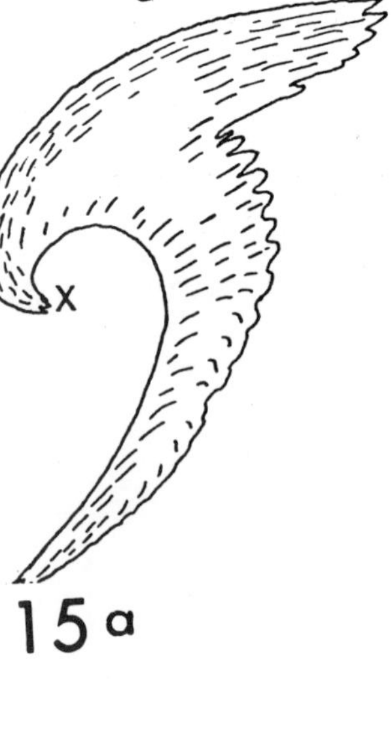

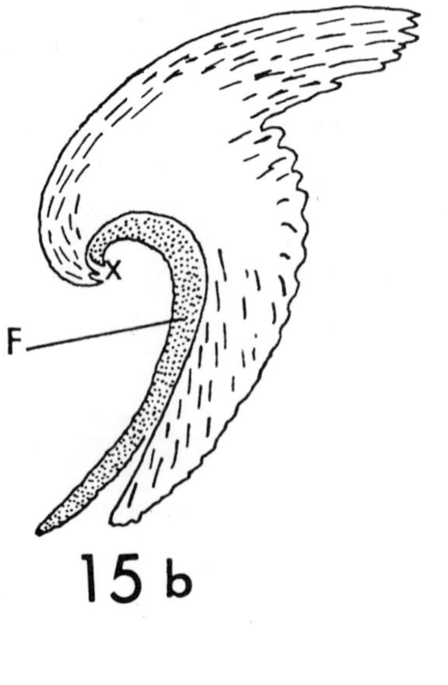

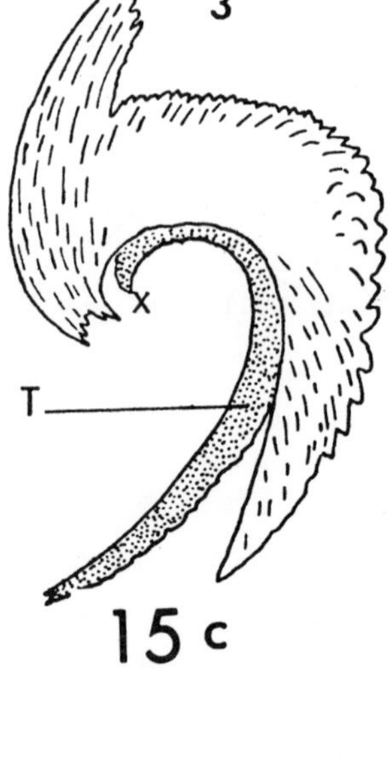

FIGURE 13

The two comma cloud patterns in Figure 13 do not have jet stream cirrus decks associated. Generally, such comma systems are located under a large upper air low or trough on the left side of the jet stream. The cloud tops of the comma in Figure 13a are middle clouds or cirrus from convective cells, that does not persist and merge with that from adjacent cells. The border on the upstream side of the comma "head" is not well defined. In Figure 13b, the upstream border of the comma head is well defined. The outflow level cirrus persists and forms an overcast. In this case, a localized "system scale" ridge has formed aloft. Such a small scale ridge directly related to the comma will be referred to as a "couplet" ridge.

FIGURE 14

Shown are three comma cloud systems with jet stream related cirrus decks present. However, the upper level flow field at jet stream height is out of phase with the mid tropospheric flow. Such patterns usually develop when the new comma formation is within the pre-existing jet stream zone or just to the left or cold side of the jet zone.

FIGURE 15

This shows three examples in which the upper tropospheric flow is nearly in phase with the mid tropospheris system. The high level baroclinic zone is also deformed into a comma shape. This condition occurs early in the developmental cycle when the new comma system forms on the right side of the old jet stream, or at the base of a deep trough with a split or branching jet stream on the west side of the trough.

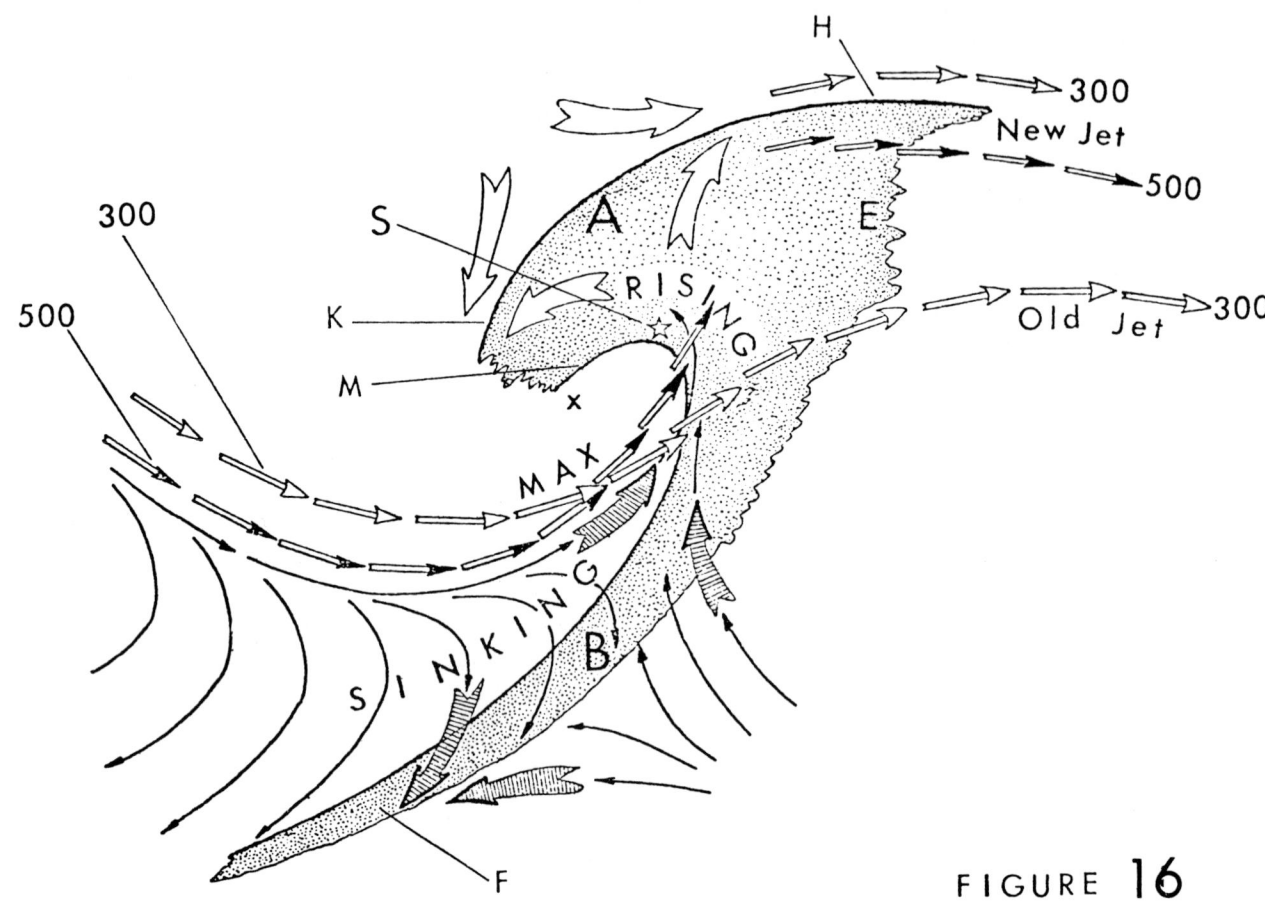

FIGURE 16

FIGURE 16

The drawing shows a model of a typical comma cloud system in the westerlies and its related wind field features. The model was empirically derived by comparing numerous comma systems on satellite imagery with surface and upper air data. The clouds involved with such a system, indicated by the shaded region on the drawing, are are not all of the same type or vertical distribution.

On the drawing, broad open arrows indicate axes of **max**imum winds at 300mb (open arrowheads) and at 500mb (solid arrowheads). Thin lines with arrowheads are surface streamlines. Very large arrows at "A" and "B" indicate hyperbolic flow in the relative motion. The deformation zone at location "A" is in the upper level flow where the air which has risen forms a distinct boundary with the upstream flow. Some of the rising air falls back with respect to the system motion and dissipates at "M"; other air accelerates downstream and dissipates its cloud elements at the axis of the upper level ridge near "E". On the downstream side of the comma "head" two axes of maximum winds at 300mb are shown. The mid and upper tropospheric air between those speed maxima is warming with time. This warming, which correlates with the amount of clouds and precipitation at the upstream side of the comma head, reduces the baroclinity with the "Old Jet" and increases it at the "new Jet". This process causes the original downstream jet stream to weaken and dissipate with time, and forms a new jet stream speed maximum to the north.

The cloud comma system in this model represents a storm in its developmental phase. A surface low pressure center would be located near position "S" and rapidly deepening. The wind fields at and above 700mb would not likely have a "closed" cyclonic circulation center, and the system would be rapidly moving.

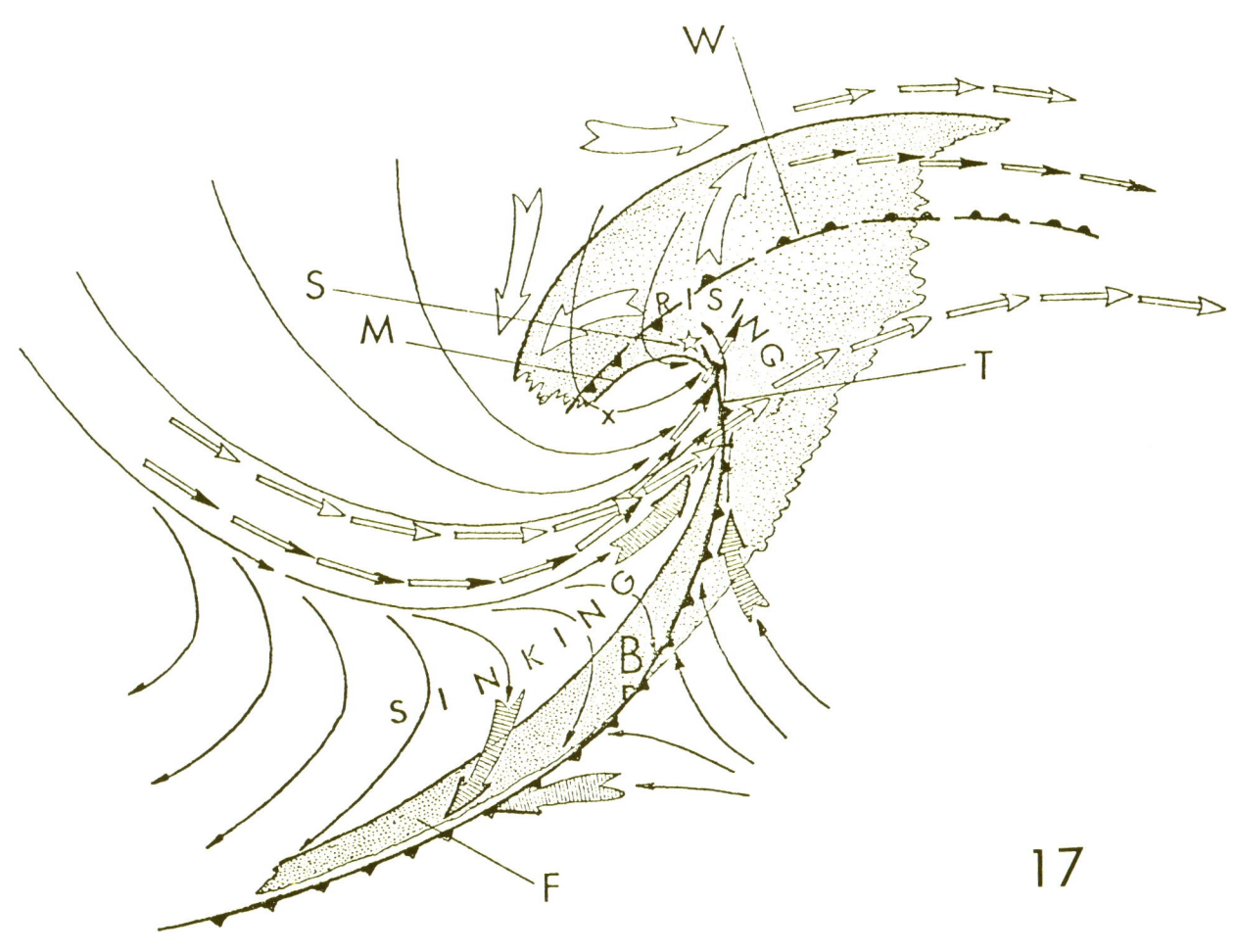

FIGURE 17

This shows the same model with surface frontal locations and additional surface streamlines. The deformation zone at "B" and the clouds of the comma "tail" feature are best associated with the low level flow field. Note that the surface cold front crosses the tail, from the back side north of the jet stream to the front side on the southern end. This is a reliable relationship. The surface warm frontal position is not as reliable. The warm frontal position vary considerably from storm to storm; and more than half of the storms do not have a warm front during their early phase of development. The presence of a warm front is dependent upon the type of cyclogenesis, and the environment in which the development occurs

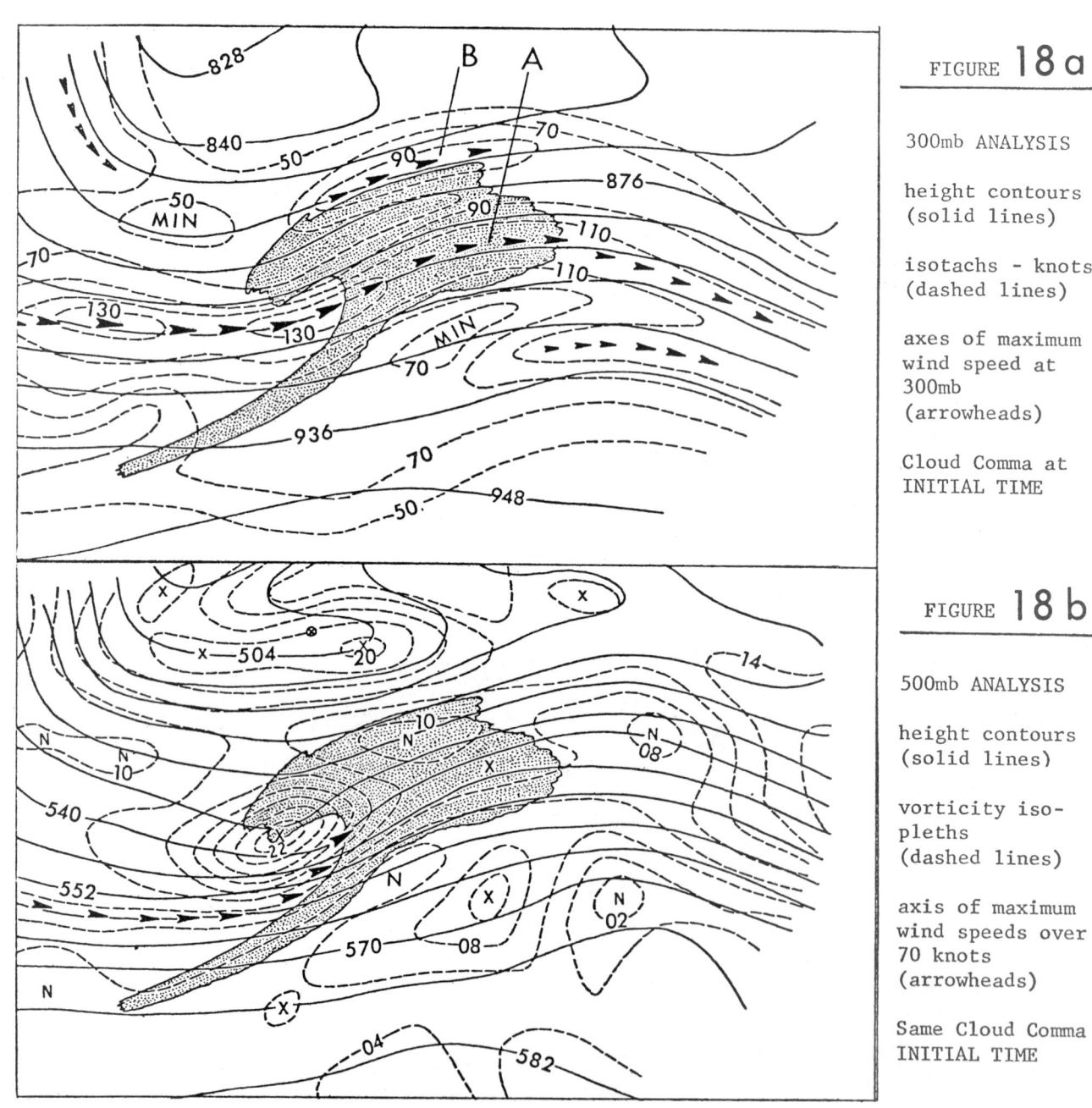

FIGURE 18a

300mb ANALYSIS

height contours (solid lines)

isotachs - knots (dashed lines)

axes of maximum wind speed at 300mb (arrowheads)

Cloud Comma at INITIAL TIME

FIGURE 18b

500mb ANALYSIS

height contours (solid lines)

vorticity isopleths (dashed lines)

axis of maximum wind speeds over 70 knots (arrowheads)

Same Cloud Comma INITIAL TIME

FIGURE 18

The four drawings of Figure 18 show a comma cloud system at two times 12 hours apart. The outline of the clouds was traced from IR imagery, and 500mb & 300mb analyses were superimposed.

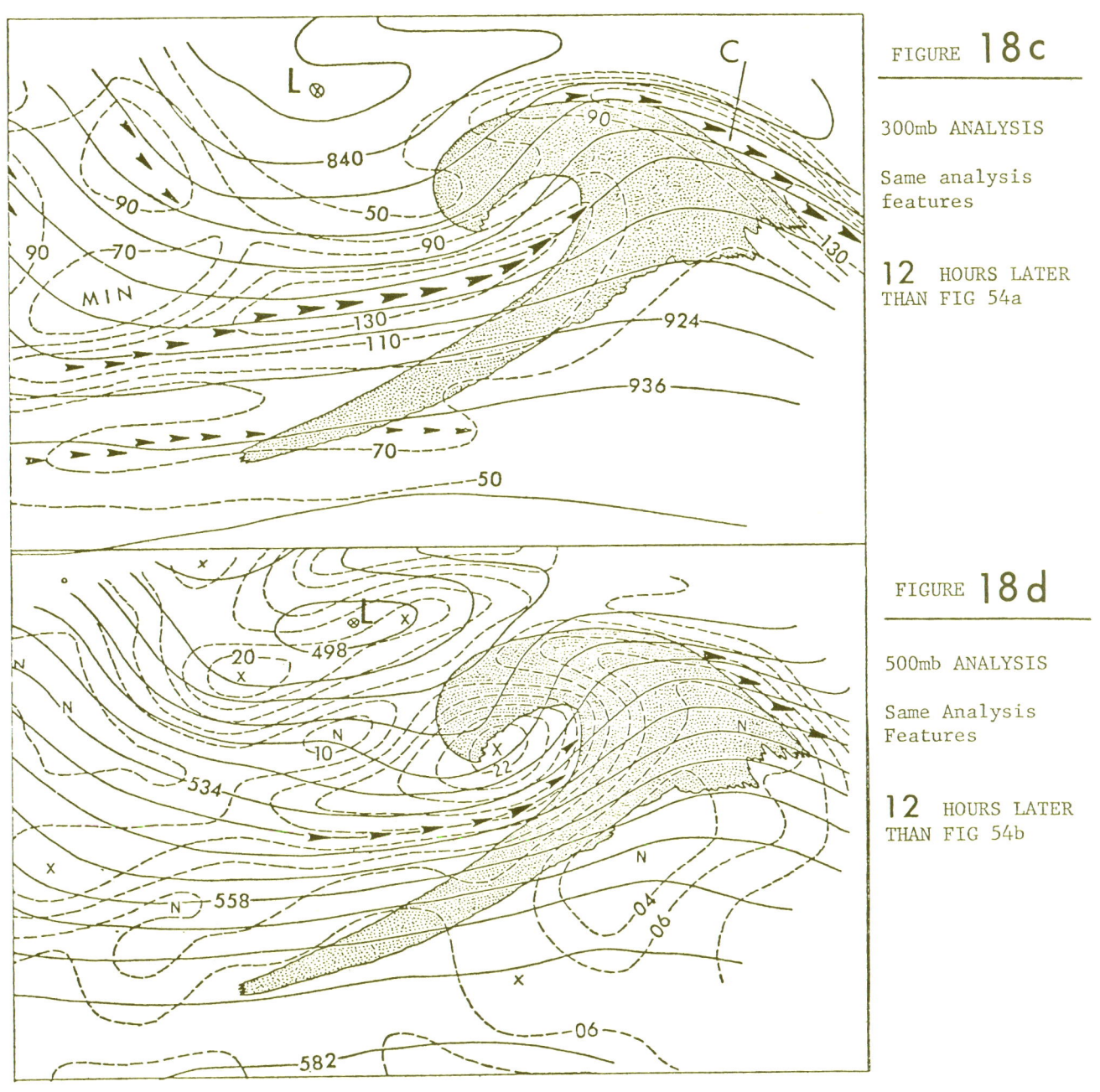

FIGURE 18c

300mb ANALYSIS

Same analysis features

12 HOURS LATER THAN FIG 54a

FIGURE 18d

500mb ANALYSIS

Same Analysis Features

12 HOURS LATER THAN FIG 54b

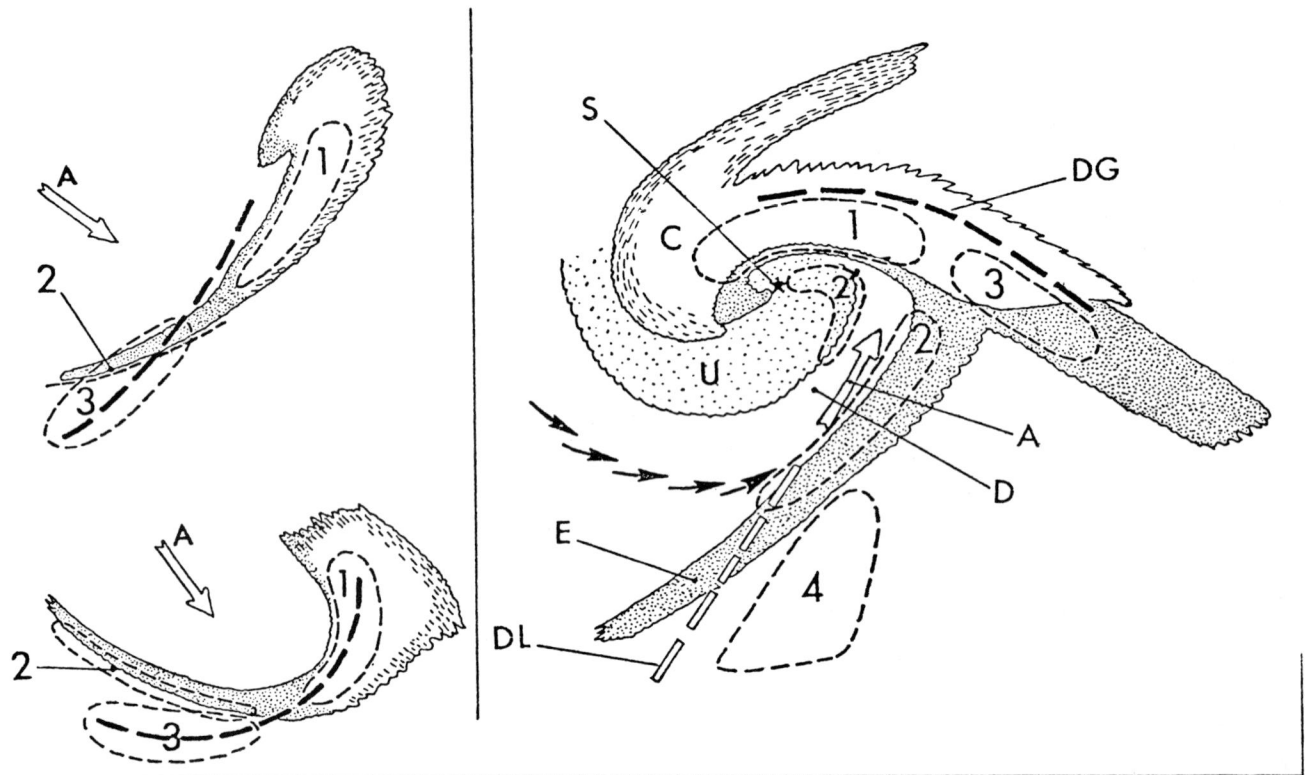

FIGURE 19

The drawings show areas of heavy precipitation observed with two types of young short wave type comma cloud systems, and one mature winter storm cloud pattern. The large open arrows indicate the position of the primary jet stream speed maximum associated with each system. The broad dashed lines show the locations of deformation zones in the mid tropospheric flow. Heavy precipitation areas are numbered and enclosed by thin dashed lines.

Precipitation Area 1 is sustained precipitation and diurnally persistent. Deep convection may be included. Area 2 is convective and has an afternoon and evening diurnal preference. Area 3 has a diurnal preference between midnight and just after sunrise. It is frequently convective, but may be sustained precipitation. Area 4 is convective and diurnally persistent, when it is present with a storm. It is more likely to be present when the upstream jet stream splits into two branches, one turning northward as shown, and the other extending to the southeast to a vorticity maximum or closed low east of area 4.

PART FOUR — THE "BAROCLINIC LEAF" CLOUD SYSTEM

"Baroclinic Leaf" is a name given to a type of elongated cloud pattern frequently present prior to the beginning of surface cyclogenesis in the westerlies. It is the cloud system associated with the "frontogenetic phase" of system development. Mid tropospheric frontogenesis has begun, new clouds and precipitation have formed; but, the surface pressure has not yet begun to lower significantly, and a single surface cyclonic center has not formed or consolidated along the surface trough or cyclonic shear zone. Approximately 75% of the leaf cloud systems observed on satellite imagery evolved to comma systems with surface cyclogenesis beginning during the transitional phase.

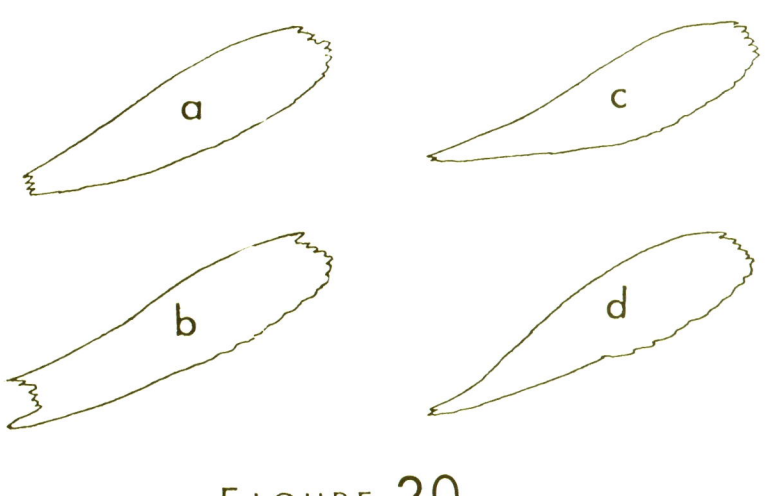

FIGURE 20

FIGURE 20

Four examples of "baroclinic leaf" cloud patterns are shown. The characteristics of the pattern are the elongated shape with both of the long side borders generally well defined. The upstream border, which is usually the western or poleward side, usually has a shallow amplitude "S" shape. The downstream border is often well defined also along its entire length, or at least along the half closest to the equator. Generally, that equatorward half of the downstream border is associated with a new or reintensified surface cold front. If the cloud system is vertically deep and low level moisture is present, significant precipitation is likely occurring. An elongated surface trough and cyclonic shear zone is likely under the cloud system and oriented nearly parallel to the cloud pattern. Some leaf systems observed on satellite imagery, often when associated with a subtropical branch of the jet stream, are limited to upper level cloudiness. Surface features are not present.

FIGURE 21

The cloud patterns shown in Figure 21 were traced from satellite imagery. The drawings show a 12 hour time sequence at 3 hourly intervals.

During the period, the leaf pattern in Figure 21a evolves into a comma pattern by Figure 21e.

When a leaf evolves to a comma pattern, it is common to have cloud dissipation at the rear half of the leaf west of the inflection point of the upstream border. Sometimes, but less frequently, the amplitude of the "S" shaped border merely increases with the western half becoming narrow to form the comma tail.

During such a transition as illustrated, surface cyclogenesis begins.

The development shown did not occur in an old frontal zone. A new cold front formed along the comma tail feature. When the development does occur in a pre-existing frontal zone, the cloud system is refered to as a "wave" on the front.

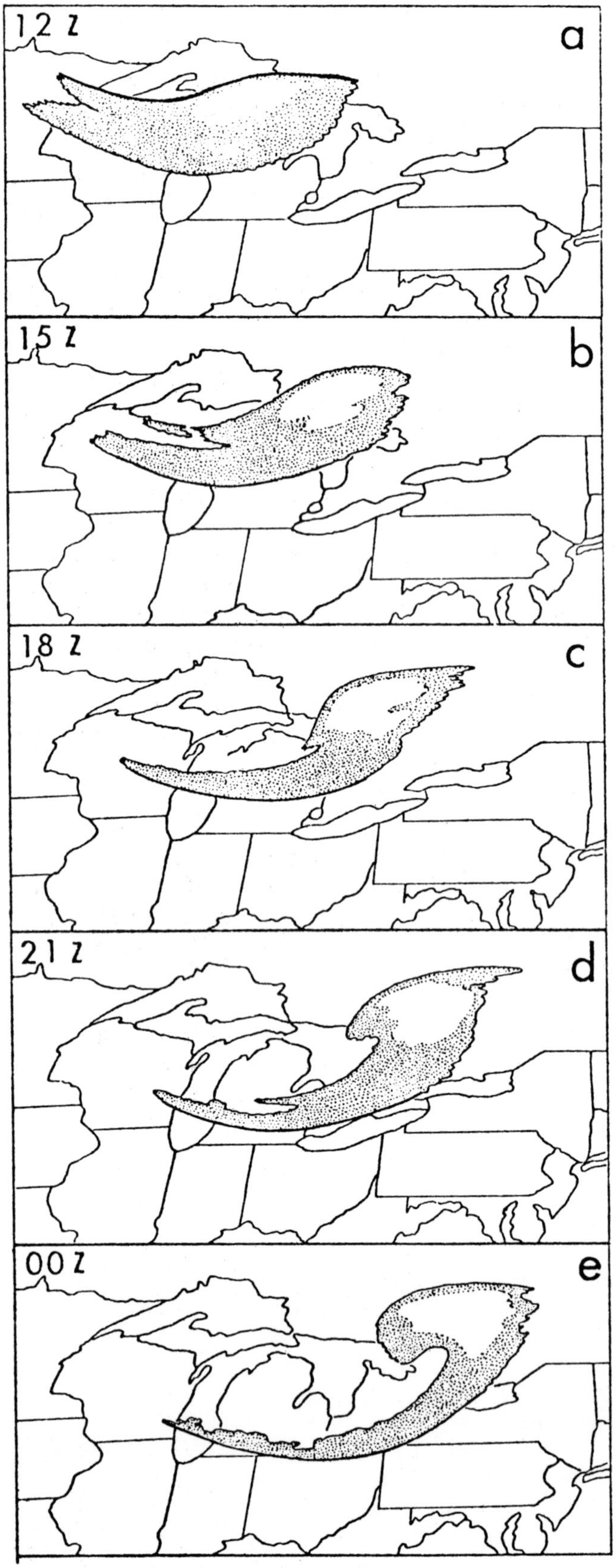

FIGURE 21

FIGURE 22

The cloud patterns on Figure 22 were traced from satellite imagery. Associated analyses and data were superimposed. The first four drawings show the same baroclinic leaf cloud pattern with different analyses. The last two drawings show the cloud systems with surface analyses for the initial time and 12 hours later.

Figure 22a has 300mb height contours and isotachs superimposed. Arrowheads indicate axes of maximum wind speed. Note that the leaf cloud system is located in the left front quadrant of a strong speed maximum. At and downstream from the leaf, strong winds are to the southeast. No wind speed maximum is located near the northern border of the cloud system. Those relationships between the leaf system and upper air winds are very common. Generally the jet stream crosses the leaf pattern near its western tip, with lower wind speeds on the downstream side. An exception to this is when a leaf forms to the right side of the existing jet stream. In that case, the downstream jet is very strong and located near the northern border of the leaf. In such cases the jet stream wind speeds can be even greater on the downstream side of the cloud system than on the upstream side.

Figure 22b shows the surface analysis with the location of the maximum 300mb winds indicated by bold arrowheads. Thin dashed lines are surface streamlines. No streamlines are shown where winds are light and variable, or over the ocean.

Figure 22c shows 500mb height contours, isotachs, and axes of maximum wind speeds. The same height contours are shown in Figure 22d with 500mb vorticity as dashed lines. Lobes of cyclonic vorticity form an "L" shape northwest of the leaf. This is common with one lobe (dash-dot line) parallel to the upstream jet stream, and another (dashed line) parallel to the leaf.

The cloud system shown developed on the cold air side of the pre-existing frontal zone, and to the left of the jet stream. A new surface cold front formed as the leaf developed, and no warm front exists.

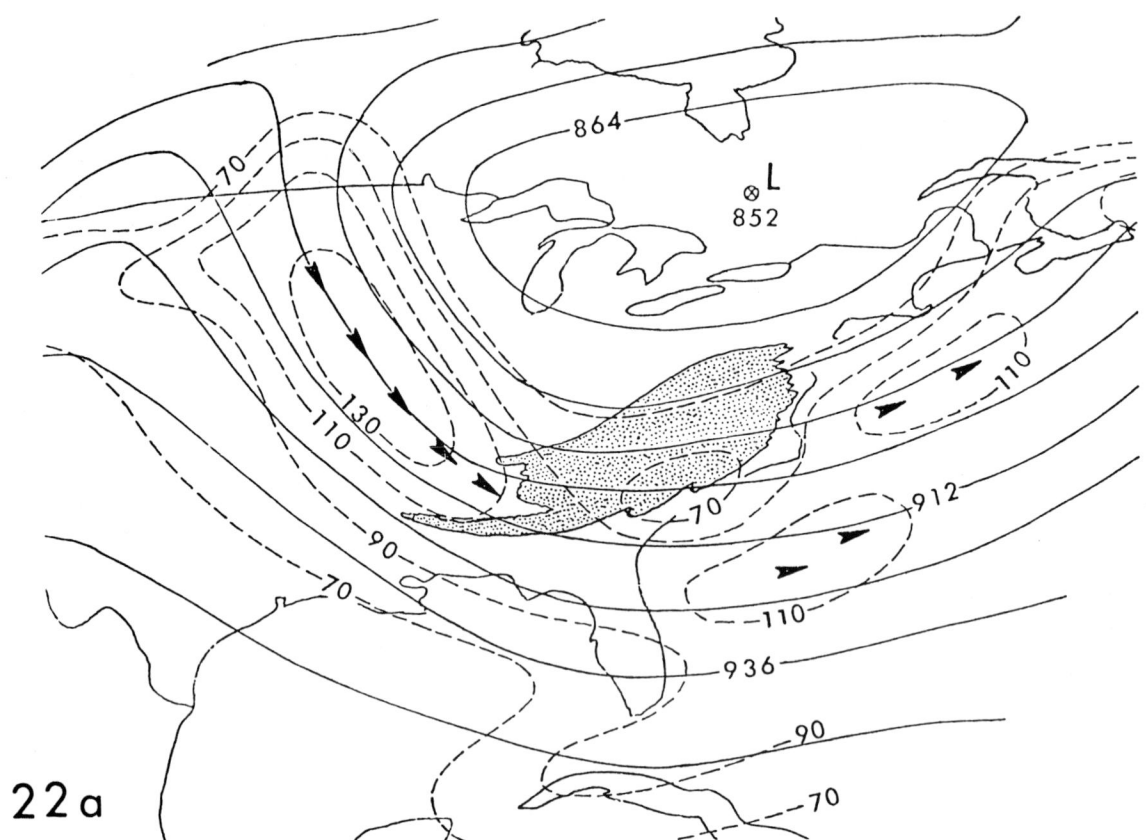

22a 300mb Analysis: Height Contours, Isotachs, Axes of Maximum Winds
Discussed on page 23

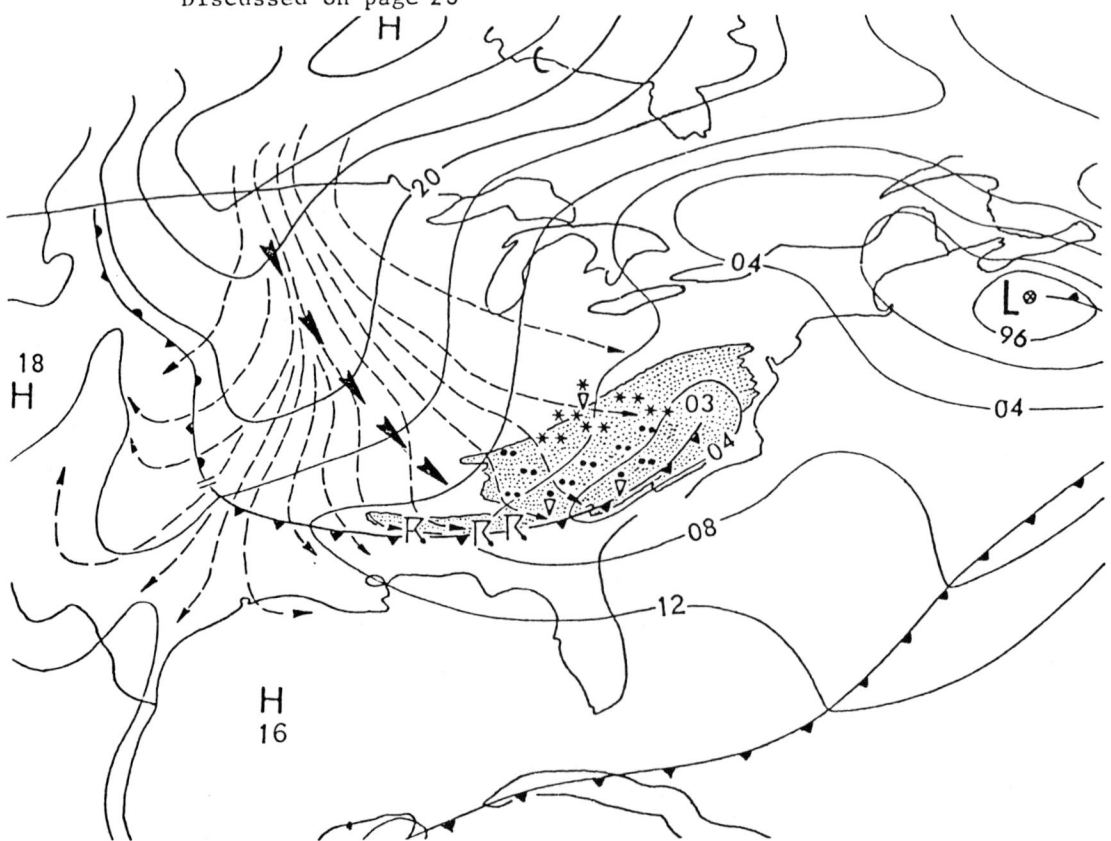

22b Surface Analysis: Isobars, Fronts, Streamlines, Precip, 300mb Max Wind Axis
Discussed on page 23

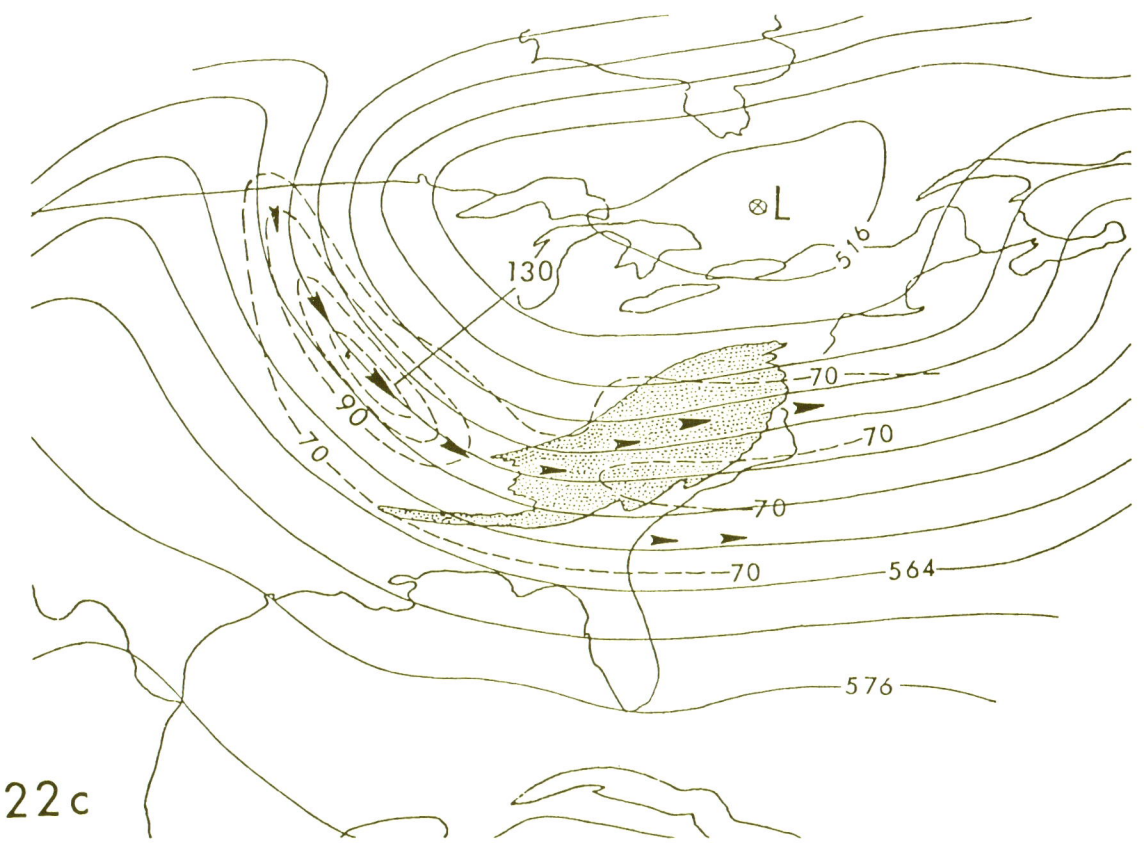

22c

500mb Analysis: Height Contours, Isotachs, Axes of Maximum Winds
Discussed on page 23

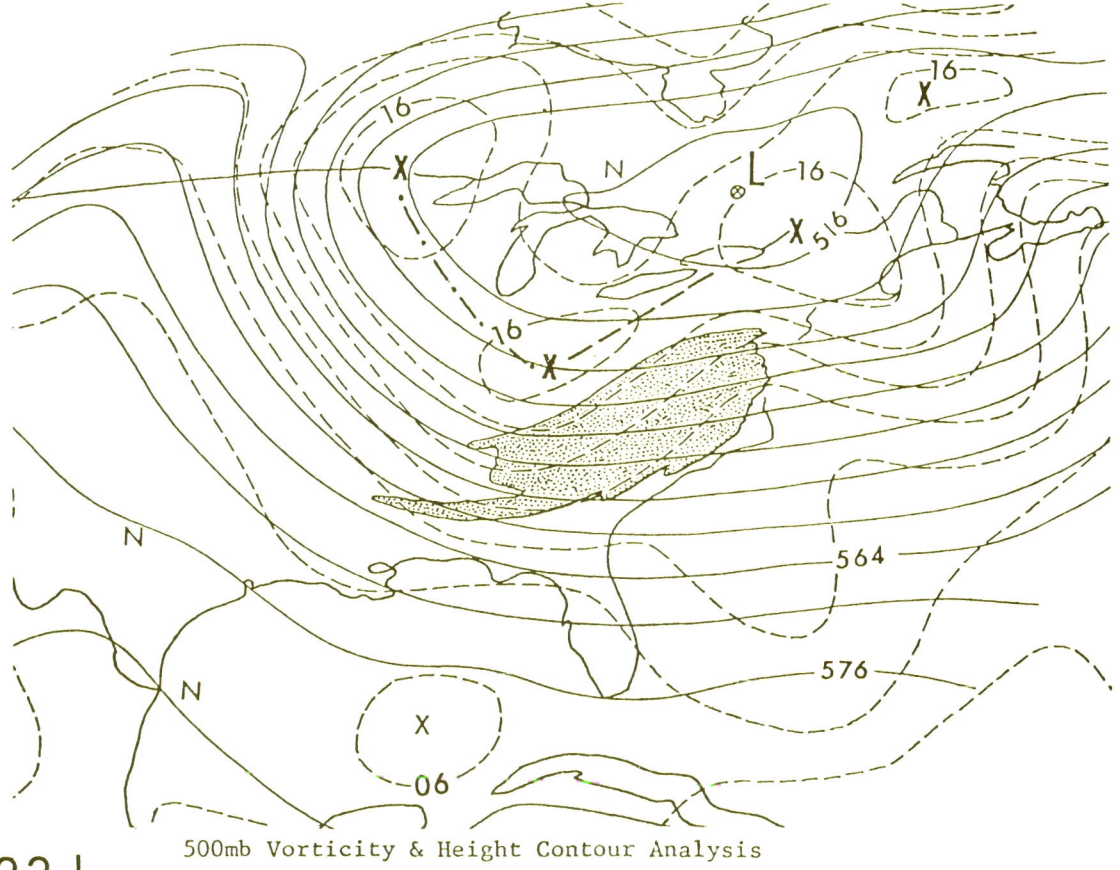

22d

500mb Vorticity & Height Contour Analysis

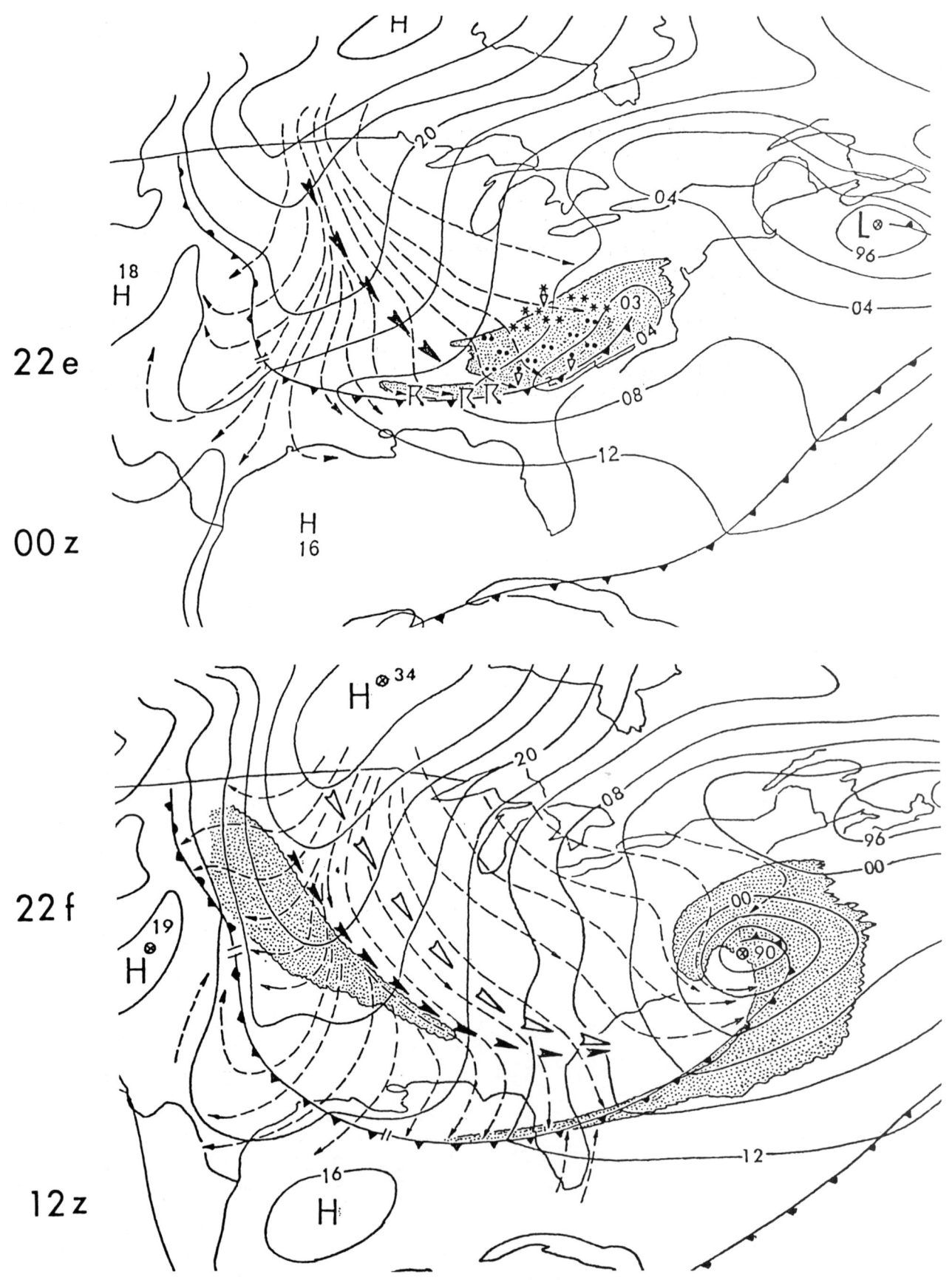

FIGURES 22e and 22f

Figure 22e is the same as Figure 22b. The initial surface analysis and cloud pattern are shown.

Figure 22f shows the surface analysis and cloud pattern 12 hours after the time of 22e. On Figure 22f, the large open arrows indicate the axis of the jet stream, and the solid arrows indicate the axis of maximum winds at 700mb.
The baroclinic leaf system has evolved to a comma pattern. A well defined surface low pressure center has formed with a central pressure of 990 mb. Note that most of the cloud pattern is at this later time offshore; precipitation is not shown. Other surface features offshore were analyzed with time continuity and ship reports.

PART FIVE   THE D.A.V.E. MODEL   AND CLOUD PATTERN EVOLUTION

After observing hundreds of storm system developments on time lapse satellite imagery, and comparing the cloud patterns to their associated wind & density fields; some fundamental common aspects have been determined, and significant differences in the storm cloud patterns have been related to differences in the wind field evolution.

Storm system developments and cyclogenesis have been empirically categorized into types and further subdivided. All of these can be considered as modifications to a basic model which I have named the D.A.V.E. PROCESS. The letters refer to: "Deformation, Advection, and Vertical Exchange". The model is summarized by the drawings in Figure 23.

FIGURE 23

A new speed maximum in the jet stream zone forms, and in advance of that speed maximum - to the left front quadrant in advance of the cold air - deep layer frontogenesis occurs by the deformation process. The related cloud & weather system is a baroclinic leaf pattern shown in Figure 23a. Vertical exchange processes begin as the new baroclinic zone forms. The upward motions in the weather region modify the upper atmosphere on the downstream side of the system, and sinking air contributes to surface cold frontogenesis at the rear of the system. The leaf pattern changes to a comma pattern. During this time rapid surface pressure falls occur near the intense weather region, and a new surface low forms. A new upper air baroclinic zone and speed maximum forms on the poleward side of the upper level warming in advance of the intense weather region.

By the time of Figure 23c, the system evolves to what I refer to as the "open comma" phase. The system is undergoing rapid surface cyclogenesis, and the upper level trough-ridge "couplet" is amplifying - especially on the downstream ridge portion. At and above 700mb, the wind field is open westerlies - no easterly flow is present. The upper air flow field is dominated by two speed maxima. These may be connected in an isotach analysis and considered parts of the "jet stream" which "shifts northward" across the comma pattern; but, it is useful to consider them separately. The upstream maximum at "A" is primarily induced by horizonal advection from upstream. The maximum at "C" is primarily induced by vertical exchange and upper air warming.

The new low level low pressure system which would be closed on the surface and 850mb analyses, is on the right side of the northern jet stream speed maximum, and to the left side of the southern one. The low has warm core characteristics on its poleward side where the east winds decrease with height, and the cloud debris falls back from the cell motions. To the equatorward side, the low has cold core characteristics; the west winds increase with height, and the cloud debris moves in advance of the cell motions.

From the open comma phase of Figure 23c the system may evolve in different ways. Two are shown in Figure 24.

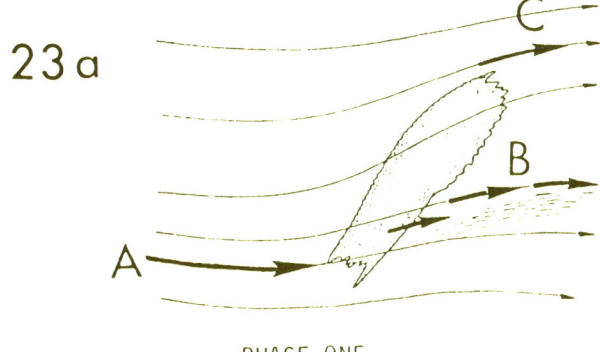

23 a

PHASE ONE

"FRONTOGENTIC" OR "ORGANIZATIONAL" PHASE

An elongated cloud system of the "baroclinic leaf" type develops. Precipitation begins. If convective weather were already occurring in a random manner in the area, it becomes organized into a linear pattern.

An elongated surface shear zone and trough have formed.

High and low level boundaries have formed as vertical exchange began.

A new upper air speed max is forming at "C" as upper tropospheric warming occurs between "B" and "C".

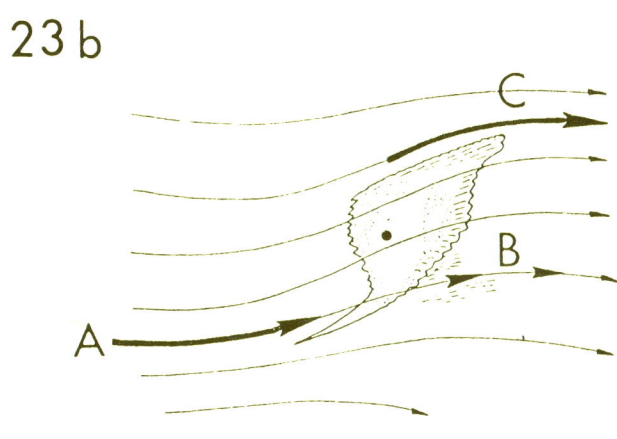

23 b

PHASE TWO "INITIAL CYCLOGENETIC" PHASE

The "leaf" pattern changes to a comma pattern; or, lines of convection form curved patterns.

A consolidated surface low forms and the central pressure begins to lower. (The lowering of pressure in advance of the low exceeds the raising to the rear) The black dot indicates the surface low position.

Westerly winds aloft decrease with time at "B" and increase at "C". This is correlated to upper tropospheric warming downstream from the heavy precip region.

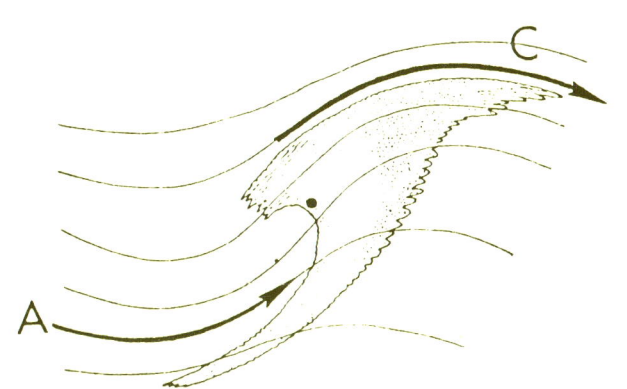

23 c

PHASE THREE "OPEN COMMA" PHASE

The surface low at the dot is well defined and deepening rapidly.

The streamlines have increased in amplitude, especially in the anticyclonic portion.

The downstream upper air speed maximum has reformed from position "B" to position "C".

24a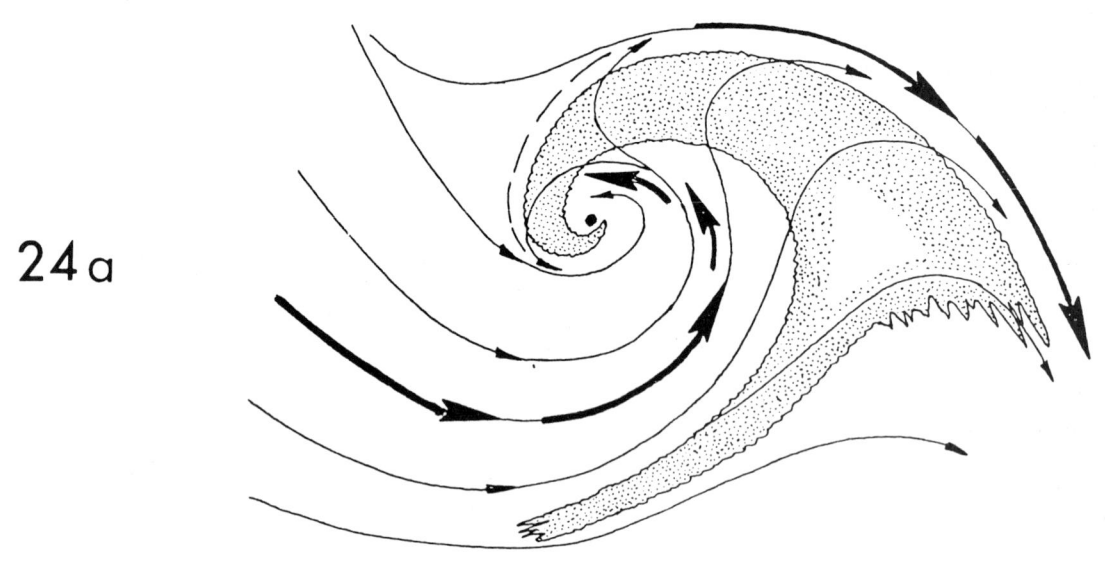

24b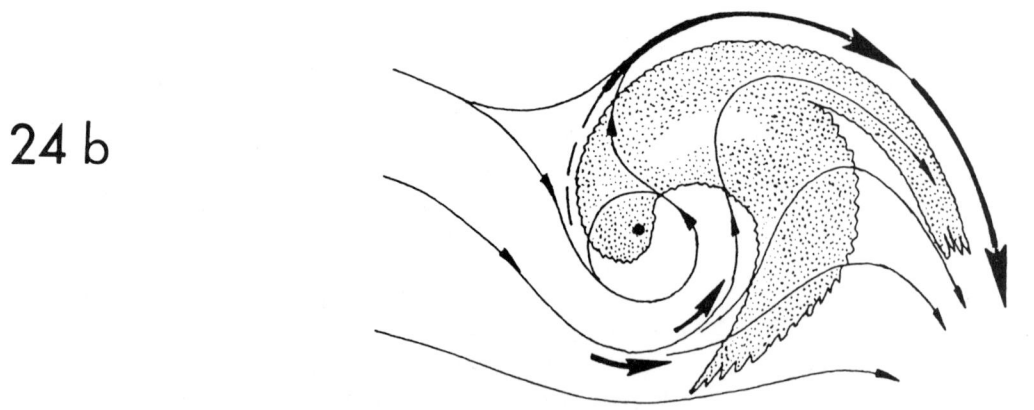

FIGURE 24

Two cloud systems and their upper air wind fields are illustrated. The shaded regions depict the "cold cloud shields" - that part of the cloud system which is both bright on VIS imagery and cold on IR imagery. This means that low topped clouds and thin cirrus would not be included.

The pattern in Figure 24a has a large concave border to its cold cloud shield with respect to its convex border.. This is characteristic of a "winter storm" pattern. Also, the region of cyclonically curved streamlines is large with a relatively narrow ridge wrapped part of the way around. The upstream speed maximum, or "inside jet", is long and strong. This speed max has extended part way around the surface low, which has rotated under the newly closed upper level low. The surface low now has winds increasing with height around its periphery, except for a small portion. It has taken on mostly cold core characteristics. The cloud pattern includes a well defined "tail" portion associated with a strong low level cold front.

In Figure 24b, the ratio of the concave border to the convex border of the cold cloud shield is much smaller than in Figure 24a. This is characteristic of subtropical storm systems that often evolve to tropical cyclones. The region of cyclonically curved flow is small, and the "inside jet" is small and weak. As the west winds weaken aloft, the surface low has taken on primarily warm core characteristics. The ridge portion of the upper air flow is relatively large with respect to the trough, and the "outside jet" dominates the speed field.

CONVECTION vs ADVECTION

When the weather occurring with the system is intense, it is frequently highly convective in nature - even with winter season systems. Therefore, I have - somewhat loosely - used the term convection to refer to all of the intense upward vertical motion. Either system shown in Figure 24 could evolve from the pattern in Figure 23c. In either case, both the advection and convection processes are occurring. If the convection process dominates, the system evolves to the 24b pattern; if the advection dominates, the 24a pattern evolves. These are empirically derived relationships. I refer to the "inside jet", the upper air low, and the comma tail as "advection induced" features; and the "outside jet", the upper air ridge, and the deepening of the surface low as "convection induced" features. Although the specific dynamical processes involved have been considered, adequate conclusions have not been reached.

In the case shown in Figure 24, the amount of convection is assumed to be similar in both examples. The difference is in the upstream advection. For the 24a evolution, the upstream advection increases or remains strong. In the 24b case, the convection dominates because the upstream advection was "cut-off" or significantly decreased.

PART SIX CUT-OFF LOW CIRCULATIONS AND CLOUD PATTERNS

A "cut-off" low is defined as a cold core upper air cyclonic circulation system which has become "cut-off" from the jet stream or speed maximum originally associated with it. Generally, this speed maximum is the polar jet stream, but the same process can occur with a subtropical jet stream, or any branch of strong upper level winds. The classical example is shown in Figure 25 for the Northern Hemisphere westerlies.

Figure 25

Figures 25 and 26 show typical long wave cycles which occur in the mid latitude westerlies. The streamlines represent upper tropospheric flow - typical of 250mb. The arrowheads indicate the location of maximum wind speeds. Figure 25 illustrates a "roll-over" cycle. The upstream ridge (R-1) amplifies and "rolls" over the norther part of the trough isolating a portion of cold air at the base of the trough. By the time of Figure 25b, the low is "closed" (streamlines encircle the low - east winds are present), but the jet stream still extends around the base of the trough. By Figure 25c, the jet stream "branches". A new branch extends across the trough north of the low, and the southern branch into the low is weakening. By 25d, the low is "cut-off". No speed maximum connects its circulation to the jet stream, and warm air of ridge "R-1" has merged with that of "R-2". The ridge has formed a "closed" high to the north of the low. By Figure 25e, The high has connected to ridge "R-2", and the upstream flow is beginning to branch. By 25f, the new southern branch of the upstream jet has entered the circulation of the closed low; it is no longer "cut-off". Cold air on the north side of the southern branch of the upstream jet will extend across the north side of the low, replacing the warm air there, eliminating the east winds, and "opening-up" the low to form an open trough as in Figure 25a.
   This discription has been simplified to highlight the most basic points. It provides a generalized model of the most typical type of cutting off process, and its complete evolution back into an open trough. The "roll-over" cycle is more common during the summer half of the year, and at lower latitudes.

Figure 26

The long wave cycle shown here is the opposite of that in Figure 25. A strong jet stream speed maximum forms on the front side of ridge "R-1" and "digs" the trough. Ridge "R-2" then "builds-back". A low closes at the base of the trough, but does not become "cut-off". Instead the high formed from ridge "R-2" becomes "cut-off" by Figure 26d. To comolete the cycle, the high merges with ridge "R-1", and the new jet stream extends around the north side of the high into the base of the trough.

ROLL-OVER CYCLE

BUILD-BACK CYCLE

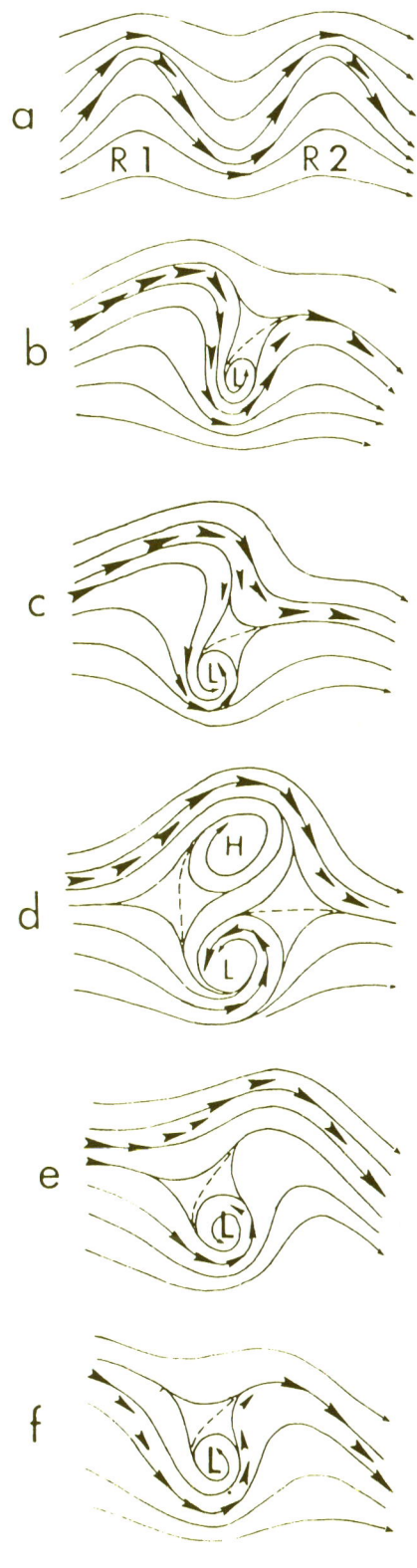

FIGURE 25

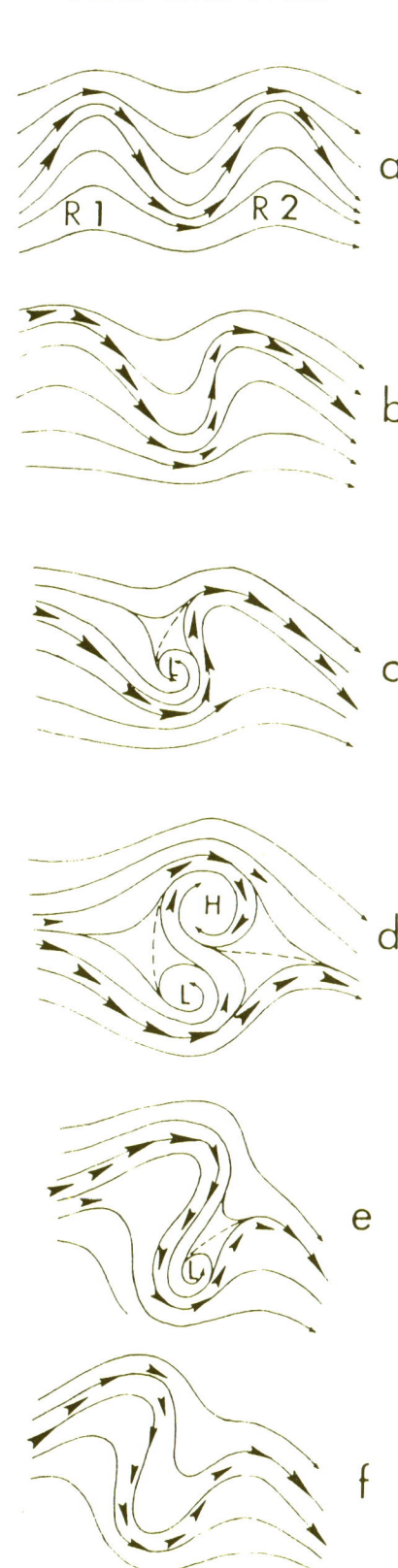

FIGURE 26

2-A-33

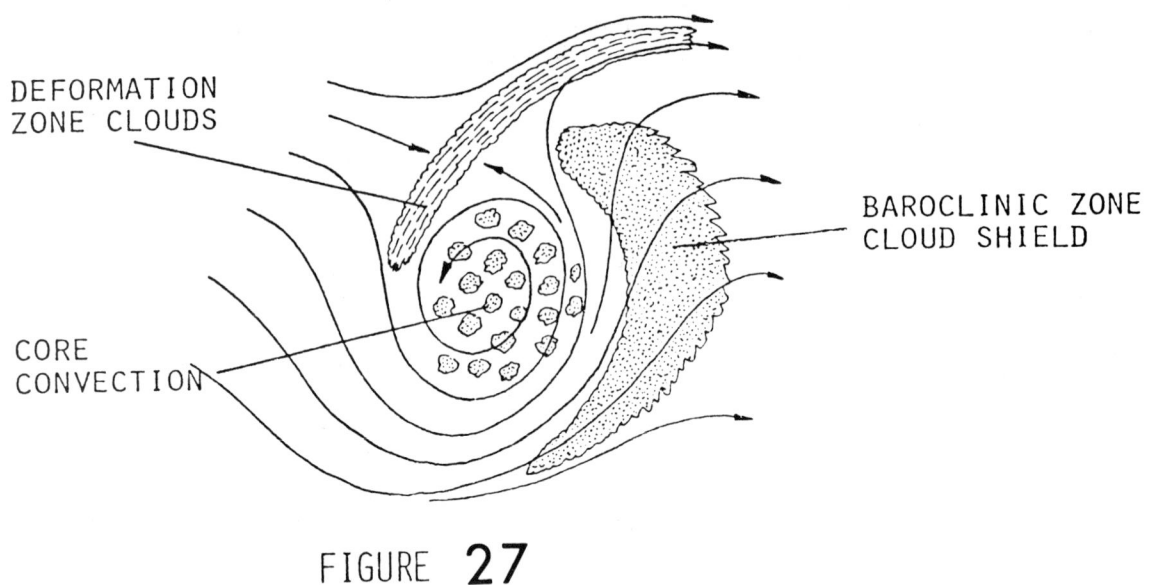

FIGURE 27

FIGURE 27

The drawing shows three types of cloudiness associated with "cut-off" low systems.

DEFORMATION ZONE CLOUDS consist of elongated bands located along the stretching axis of the hyperbolic zone. Usually these clouds are stable layers of cirrus and altostratus. Gravity waves moving thru the clouds in the direction orthogonal to their orientation may appear as striations on a single satellite image. Generally, precipitation does not occur with this type of cloudiness. If the adjacent upper air ridge forms a closed anticyclone, there will be two deformation zones involved, and both may have clouds associated.

The BAROCLINIC ZONE CLOUD SHIELD is located between the cyclonic circulation of the cold low and the axis of the "couplet ridge". The "couplet ridge" is not the nearby long wave ridge; it is the local ridge directly associated with the cut-off low. It is of the same scale as the low and is usually wrapped partly around the low. This cloudiness usually has high cold cirrus tops, and is of such vertical depth to appear very cold on IR imagery; but it may not extend down into the lower atmosphere or produce precipitation, if the upper circulation is shallow. Commonly, however, the circulations do extend thru a deep layer of the atmosphere, the clouds are low based with embedded convective cells, and significant precipitation occurs.

CORE CONVECTION occurs within the cyclonic circulation under the cold air core of the low, and consists of towering cumulus or thunderstorms. The anvils move in advance of the cell motions, and do not persist or merge with debris from adjacent cells. The presence of core convection depends upon the availability of warm moist air at low levels. It is most common over warm ocean waters during the summer half of the year.

FIGURE 28

The formation processes of "cut-off" lows can be usefully categorized into three basic types. These are illustrated in the drawings during the phase in which the low becomes "closed".

BASIC ROLL-OVER

The upstream jet does not extend beyond the ridge axis. No jet stream speed maximum extends into the low circulation. The low is "cut-off" as soon as it becomes closed. A cut-off low formed this way is least likely to have clouds associated. Deformation zone clouds are the most likely. If the system is over low level unstable air, core convection may form.

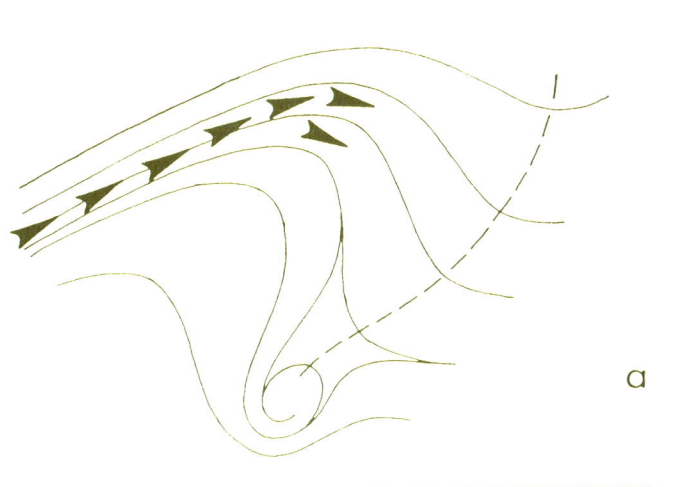
a

HOOK AND ROLL

The jet stream "hookS" anti-cyclonically into the base of the trough, but does not cross the trough. A low closes, and the poleward extension of the trough weakens or becomes a cyclonic shear zone. A baroclinic zone cloud shield is very likely with this type of development. Deformation zone clouds are also likely, and merged with the baroclinic shield.

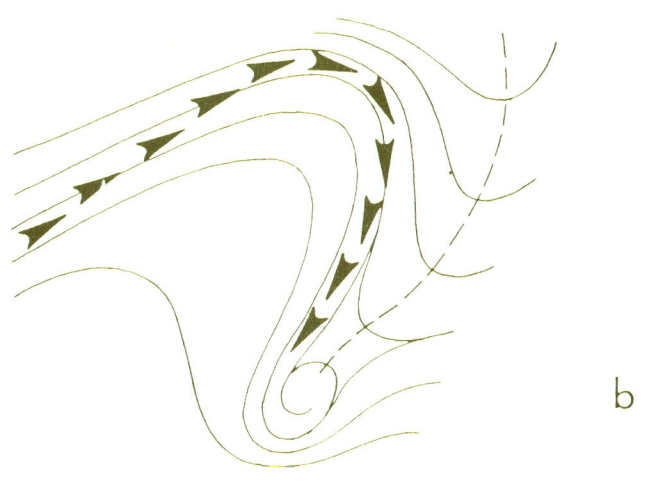
b

BRANCH AND ROLL

This is the most common type of "cut-off" development. The upstream jet branches, usually in time sequence. A speed max extends to the base of the trough and a second moves to the left crossing the trough. The first then weakens and dissipates as warm air rolls across. Significant deformation and baroclinic shield clouds usually occur, and earlier than the hook & roll formations. There is more variation of clouds & weather with this type than with the first two processes.

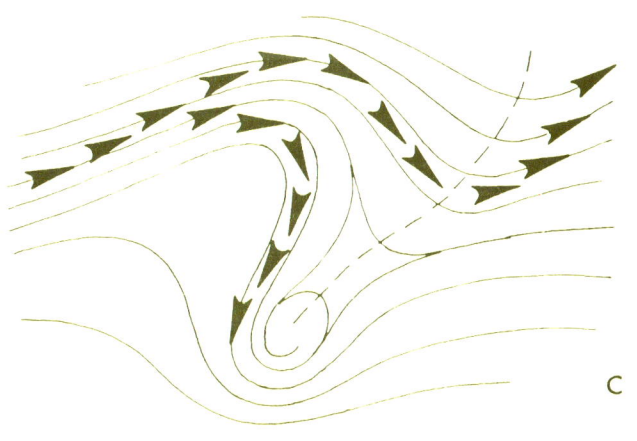
c

Figure 28

EXCERPTS FROM
PART IV CLOUD PATTERNS AND THE UPPER AIR WIND FIELD
Roger Weldon, March 1979

Jet Stream Structure: "Channel Jets" & "Advection Jets"

The "jet stream" can be thought of as some continuous belt of strong upper level winds encircling the world, which meanders meridionally in wave motions; or as several such belts with names like: "polar jet", "subtropical jet", or "arctic jet". An alternative approach is to consider these strong middle and upper level wind zones as separate (but interrelated) jet stream segments or "speed maxima", whose change with time (formation & dissipation, intensification & weakening) is of nearly equal importance as their movment. This second approach is much more useful for understanding & interpretating the behavior of cloud systems observed on satellite pictures.

If we use this second approach - thinking of the "jet stream" as an array or group of "jet segments" or "speed maxima" - then, it is also very useful to categorize the jet segments into at least two different types. I call these two types: "channel jets" (or jet channels or merely channels) and "advection jets". By definition, the difference depends on the relation between the height contours and vorticity isopleths on the 500mb surface.

If - along the zone of strongest winds or height gradient - the vorticity isopleths are parallel to the contours, the maximum wind zone there is called a "channel jet" or is "channeled".   If - in the strong wind area - the contours are NOT parallel to the vorticity isopleths - but cross at large angles, the max wind zone there is called an "advection jet".

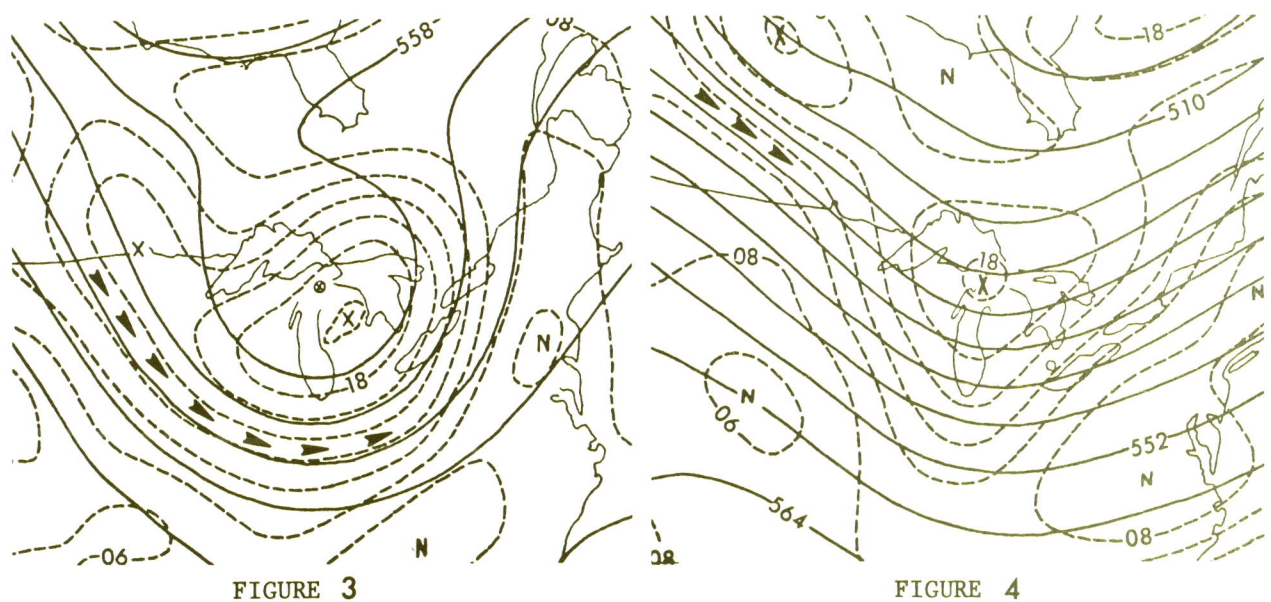

FIGURE 3   FIGURE 4

Examples of the two types of jet structure are shown on 500mb analyses in Figures 3 and 4. Solid lines are height contours and dashed lines are vorticity isopleths. Arrowheads have been placed along the strong wind zone where it is "channeled". The vorticity maximum in the center of Fig 3 has a "channeled" zone around its southwestern quadrant. The strong wind zone in Fig 4 is mostly of the "advection" type, except in the northwest corner of the drawing.

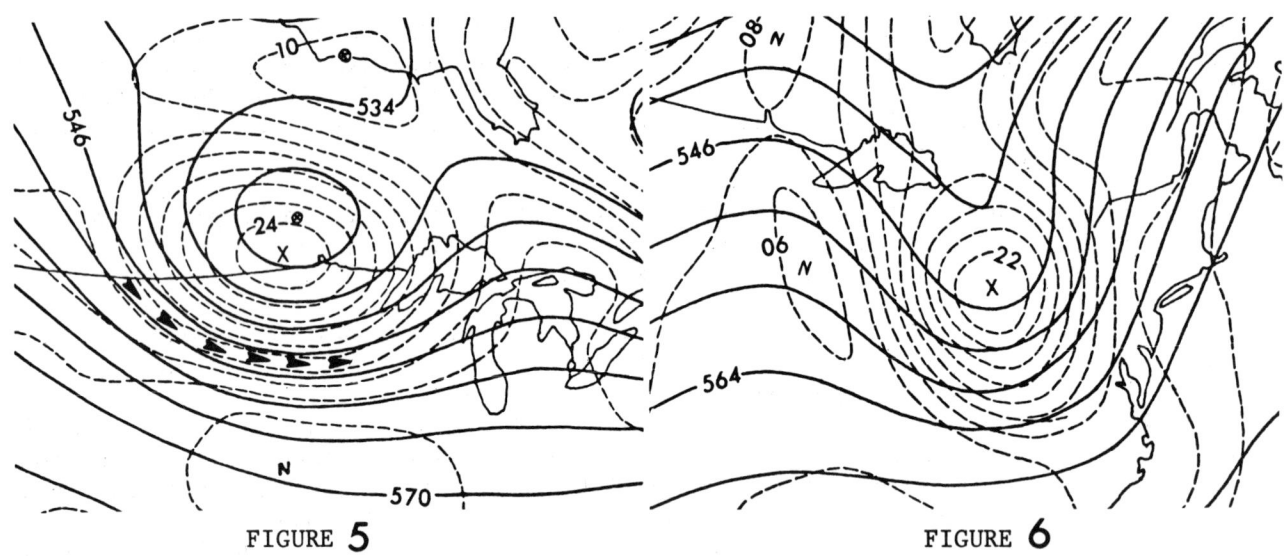

FIGURE 5   FIGURE 6

In Figure 5, a strong well defined "channeled" jet segment south of the vorticity maximum is identified by arrowheads. The vorticity maximum in Figure 6 has no "channeled" region in its strong wind zone.

Two more examples are shown in Figures 7 and 8 below. There are several vorticity maxima in each figure.

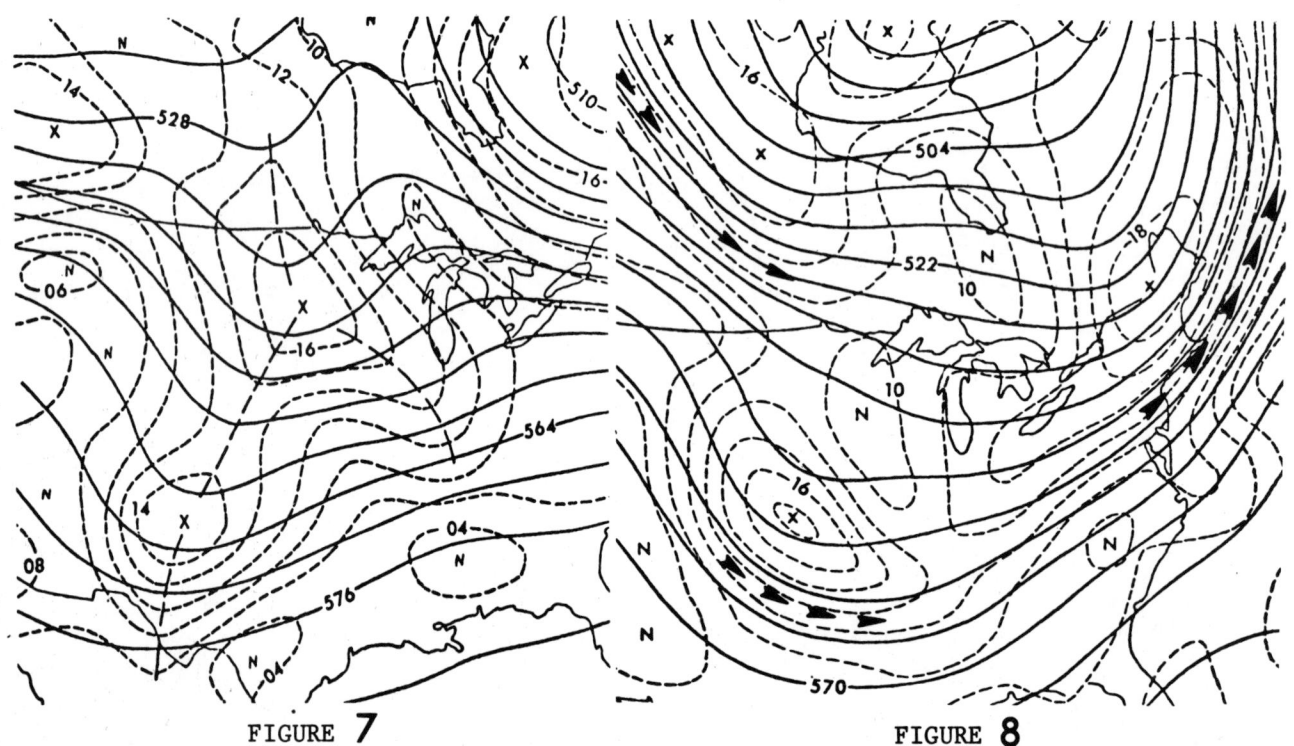

FIGURE 7   FIGURE 8

In Figure 7, the maximum wind zones are primarily of the "advection" type. In Figure 8, several "channeled" zones are present (identified by arrowheads).

Some characteristics of "channeled" and "advection" type wind maxima are listed on pages 6 & 7.

Vorticity "Lobes"

A vorticity "lobe" is an elongated ridge or trough in the vorticity field. Use of the terms "vorticity ridge" or "vorticity trough" is confusing (A "vorticity ridge" is usually found with a "trough" in the pressure or height fields). So, I will use the word "lobe", which I first heard used in this purpose by meteorologists at the NESS San Francisco SFSS.

An "advection lobe" is a lobe of vorticity whose axis crosses the height contours or streamlines at large or significant angles. (See lobe "E-F" of Figure 10 and the lobes identified by dashed lines in Figure 7.)

A "shear lobe" is a lobe of vorticity whose axis is parallel or nearly parallel to the height contours or streamlines. In Figure 9, lobes "A-B" and "C-D" are "shear lobes" maximum vorticity; they are on the cyclonic shear side of maximum wind zones. Lobe "1-2" is a "shear lobe" of minimum vorticity on the anticyclonic shear side of a strong wind zone. Note that it is the AXIS of the "shear lobe" which is parallel to the contours - not necessarily the vorticity isopleths. The axis of "shear lobe" "J-K" in Figure 11 is nearly parallel to the contours, but the vorticity isopleths cross at significant angles in a "positive advection" sense. In general, however, "shear lobes" are associated with channeled type maximum wind zones.

A "downstream shear lobe" is that portion of a shear lobe which extends downstream from the largest vorticity value (the vorticity center). In Figure 9, that part of lobe "A-B" downstream from "S" is a "downstream shear lobe", and all of lobe "J-K" (Fig 11) is downstream from the vorticity max. An "upstream shear lobe" extends upstream from the vorticity center.

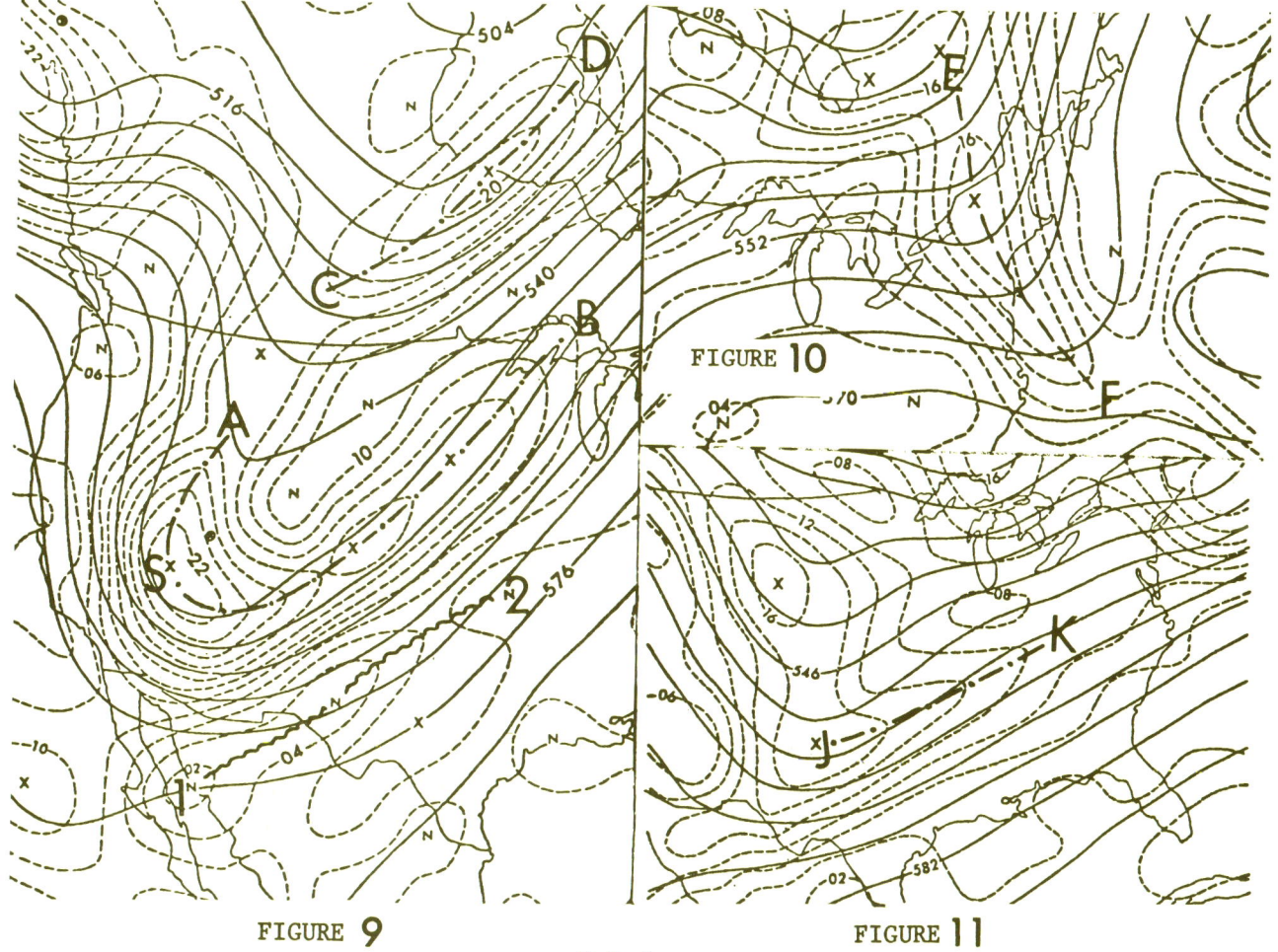

FIGURE 9   FIGURE 11

The following observations regarding jet stream structure are supplemental to the notes. The information is useful, but not essential, to understanding the remainder of the notes.

CHARACTERISTICS OF THE "CHANNEL" & "ADVECTION" TYPES OF JET STRUCTURE

1. Where the jet stream is in the "CHANNELED" configuration:

    a. Vorticity lobes are oriented parallel to the streamlines or height contours. (At 500mb)

    Generally, well defined shear lobes are associated with channeled jet structure. Vorticity advection is often present, but not along the associated maximum wind zone.

    All "jet channels" do not have shear lobes present. (This is partly dependent upon the grid interval of the analysis model and upon the density of upper air data.)

    b. The axis of maximum winds at 500mb, and usually the jet stream axis also, is parallel to the streamlines and to the height contours.

    The axis of maximum winds on a given upper air chart - such as 300mb - will usually parallel the same height contour thru the channeled region.

    c. The jet stream axis of maximum winds tends to be better defined over channeled regions (and is even better defined when a vorticity shear lobe is also present).

    d. The overall strongest winds aloft are usually found in the channeled regions as opposed to the advection structured regions.

    e. Within all well defined channeled regions, the 500mb isotherms are parallel to the streamlines, height contours, and the axis of maximum winds. During the winter season in mid latitudes, well defined channeled regions have large vertical depth. In those cases: the 1000-500mb thickness isopleths are also parallel to the axis of maximum winds, and there is little change of wind direction with height within the channeled region.

2. Where the jet stream is of the "ADVECTION" configuration:

    a. Vorticity lobes are not parallel to the winds, but cross the streamlines and height contours at significant angles

    b. The axis of maximum winds is NOT PARALLEL to the wind direction, nor to the contours.

    - Over PVA areas: the axis of maximum winds on a given upper air chart shifts to lower height contours in a downstream direction. (usually a poleward shift)

    - Over NVA areas: the axis of maximum winds on a given upper air chart shifts to higher height contours in the downstream direction. (usually an equatorward shift)

    c. The jet stream axis of maximum winds is not as well defined over regions of "advection jet" structure, as it is over "channeled" regions.

3. In general, channeled jet structure is more often associated with higher amplitude systems (troughs & ridges); and advection structure with lower amplitude systems. In other words, a closed low (aloft) is more likely to have channeled jet structure than an open wave is. But this isn't always so - see for example figures 7 & 8. The cloud patterns and their behavior are better related to the jet structure than to the wave amplitude.

4. Since the strongest - best defined - maximum wind zones are found in the regions of "channeled" jet structure, and the weaker - less well defined - segments of the "jet stream" are found in the "advection jet" regions; we could consider the "jet stream" as consisting of a series of "speed maxima" in the channeled regions connected by advection jet segments.

    The classical "speed maximum" (such as the one depicted in the center of the drawing of Figure 2 of these notes) is a jet stream segment of the "channeled" type.

GENERAL RELATIONSHIPS BETWEEN CLOUD PATTERNS & JET STREAM AXES

In this section, I will discuss some general relationships between jet stream axes and cloud patterns observed on still pictures (visible or infrared images). More information - and in quantitative format - can be obtained from satellite data by using other techniques, such as measuring cloud element motion in time sequence images, or by measuring density gradients. The objective of using the relationships presented here, is to be able to qualitatively locate the axes of the jet streams. If synoptic scale accuracy of 100% is defined as the kind that can be obtained over the U.S. upper air network using operational real time analysis techniques; then, I would say that we can locate the "jet stream" axes from still pictures with about 75% to 85% accuracy.

The task of doing this can be complex - the difficulty depends upon the amount of detail required, and upon the type of atmospheric environment involved. There are many variations that can fool even the most experienced observer. I will discuss many of these variations in following sections of the notes; but, first, it is useful to present some generalized relationships.

If I were asked to provide a simple technique of locating the jet stream axes from still satellite pictures, I would suggest using the following 4 "rules".

1. When a well defined poleward side border exists along a large cirrus deck, the axis of a jet stream is likely to be located along the border and just to the clear side of the edge. (The relationship of such cirrus borders to the jet stream was reported by Oliver, Anderson, and Ferguson (1964), and has been discussed in various literature since that report.)
2. Cloud pattern borders or edges of cloud decks formed ACROSS the jet stream, move faster at the jet axis - the wind maximum produces a bend or apex in the cloud border at the jet axis.
3. Low and/or middle clouds are often distinctly different on opposite sides of the jet stream axis. (This type of relationship was pointed out by Oliver (1968)) And the nature of cloudiness along a low level frontal zone is related to the location and character of the upstream jet stream.
4. The location of the jet axis can be interpolated between locations found by rules 1, 2, & 3.

Each of the 4 "rules" or relationships listed above is discussed and illustrated in the remainder of this section. The rules are based upon generalized relationships. There are many exceptions and other variations. Many of these variations are systematic and repetitious; they involve specific cloud patterns and related jet stream characteristics, and will be illustrated and discussed in later sections of the notes. An expanded discussion of the above 4 rules follows:

1. Cirrus cloud decks tend to form or persist on the anticyclonic shear side of the jet stream axes - with a well defined cloud border along the axis.

Axes of maximum winds <u>aloft</u> act as a border between dry air on their cyclonic shear sides and moist air on the anticyclonic shear sides. This high level moist to dry boundary is often very distinct & well defined - even when high clouds are not present. Thus, if we observe a cloud pattern with high cold cirrus tops and a well defined poleward side border; it is probable that the axis of a jet stream is located parallel to the border - just on the clear side.

Two examples of such cirrus decks are shown in the drawings of Figure 12 below.

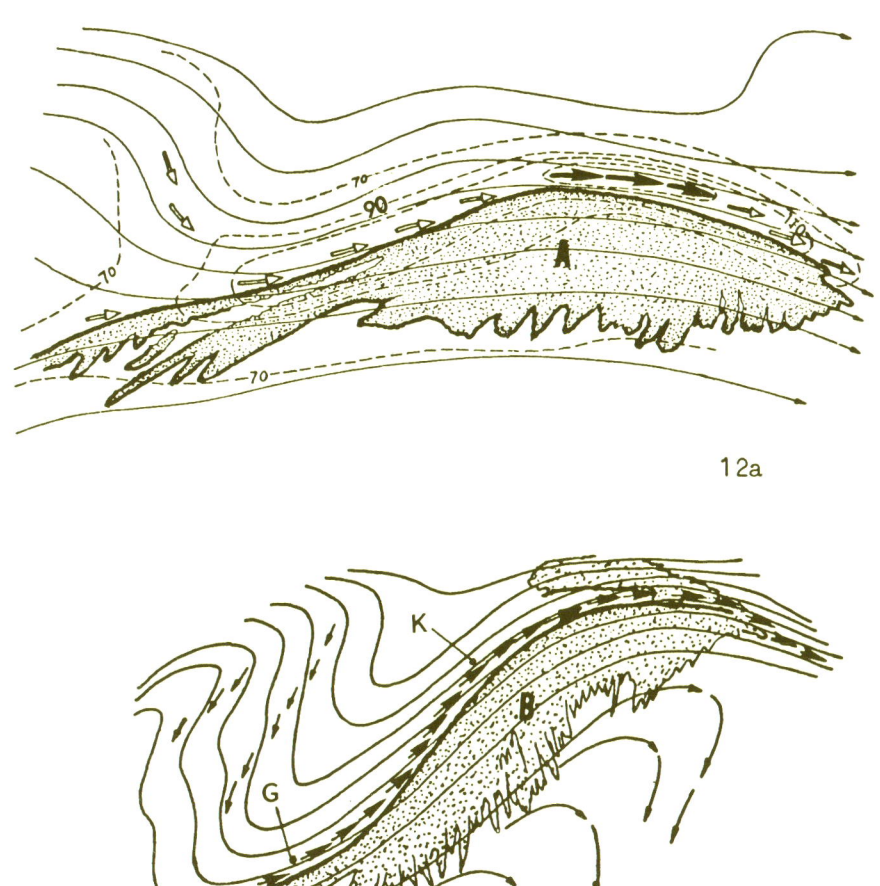

12a

12b

Figure 12:

In both drawings, thin solid lines represent 300mb streamlines; short arrows depict the location of maximum 300mb wind speeds; shaded areas are cloud patterns with high cold IR tops.

The thin dashed lines on Fig 12a (top) are isotachs at 20 knot intervals.

On Fig 12a - along the jet axis, the greastest wind speeds (over 150 kts) are shown by solid arrows. In that region, the jet is channeled, the axis of max winds and the cloud border are parallel to the wind direction.

These are common types of very large "baroclinic zone" cloud systems where the relation between the cirrus border and the jet stream axis is a good rule. Cloud system "A" of Fig 12a is primarily located over a low amplitude upper level ridge with the jet structure channeled or nearly channeled along most of the cloud border. This synoptic location - an anticyclonically curved & channeled jet stream - is the most common. The second most common synoptic location for cirrus borders along the jet axis is on the front side (east side) of a high amplitude trough, such as depicted on Fig 12b as cloud system "B". Note that along the cloud border between points "G" and "K" the axis of maximum wind is not parallel to the wind direction and that the streamlines cross the border and the axis. The jet structure there is of the "advection (PVA)" type. Dowstream from "K" over the ridge, the jet is channeled. At times cirrus borders are found along the jet axis at the base of troughs and on the west (back) side of troughs - even with high amplitude ones. These cases (not depicted) are much less frequent than on the front side of troughs and over ridges; and when it happens the jet is in the channeled or nearly channeled configuration.

Rule number 1 can be used to identify or locate the jet stream axes about 25% of the time. Or - to put it another way - of the identifiable well defined jet stream axes about 25% will have cirrus on the anticyclonic shear side with a well defined border along the jet axis. When satellite moisture channel data data becomes routinely available in a useful format, the above percentage will significantly increase. (My guess is that it will double)

Not all cirrus decks with well defined borders or cirrus bands are located along the jet axis. There are two types of cloud patterns that occur frequently, and are exceptions to Rule 1. One type is called (by me) a "baroclinic leaf", and the other is a "deformation zone" band or boundary. These will be discussed later in the notes.

The cirrus decks found to the right side (northern hemisphere) of jet axes are called (by me) <u>baroclinic zone cirrus</u>. This type of cloud pattern and Rule 1 will be discussed in more detail in Section C of the notes. (This is Sec B)

2. Where "baroclinic zone cirrus" is not present, but other high or middle level clouds are, cloud bands and boundaries will be most advanced downstream where the jet stream axis crosses them.

An illustration of this concept is shown in Figure 13 below.

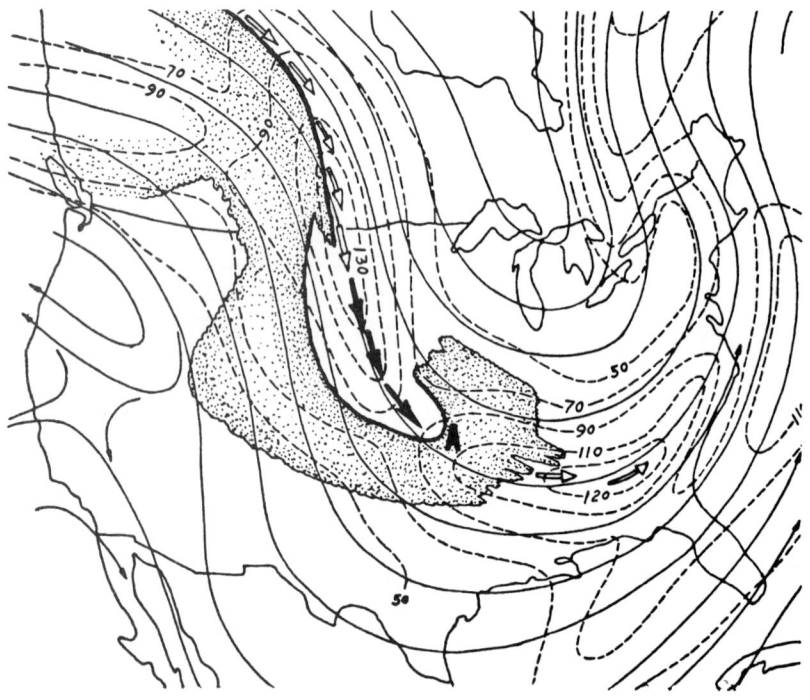

Figure 13

The drawing shows 300mb streamlines (thin solid lines) and isotachs (dashed lines) with the cold cloud area shaded and the axis of maximum winds marked by arrows.

Note location "A". The clearing zone just to the rear of "A" is most advanced downstream under the axis of maximum winds.

Frequently cloud bands and borders (especially on the upstream side where the borders are usually most well defined) form a "U" or "V" shape with the axis of maximum winds in or over the slot.

This situation usually occurs where the jet stream is NOT channeled, and is common with the cloud patterns of short-wave scale systems - especially when the system is in a long wave zonal pattern or on the front side of a long wave ridge.

Details will be discussed in Section E of the notes. Rule 2 allows us to locate the axis of maximum winds another 25% of the time.

3. When no high clouds (with tops at jet stream levels) are present, the axis of the jet stream will often be revealed as a boundary or interface between different types of low level cloud cover.

An example of this is depicted by Figure 14 below.

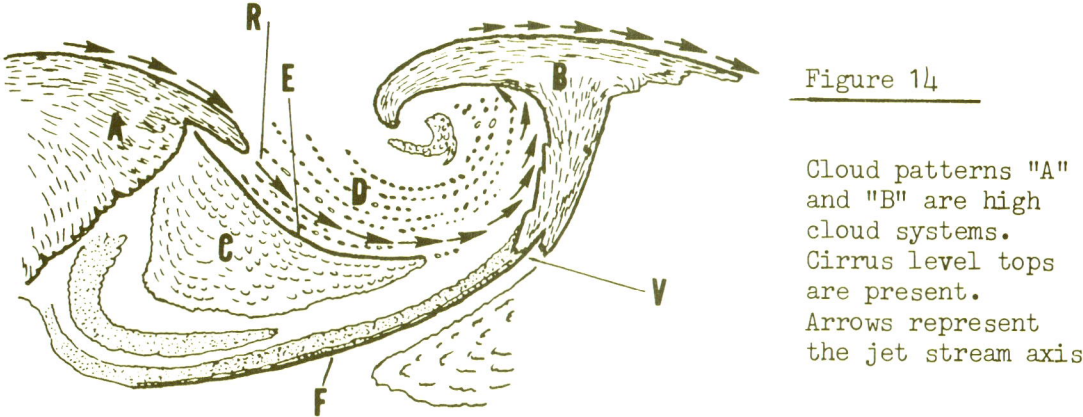

Figure 14

Cloud patterns "A" and "B" are high cloud systems. Cirrus level tops are present. Arrows represent the jet stream axis

On the drawing, line "E", which is parallel to the axis of maximum winds aloft, is an interface between two different types of lower level clouds. Such interfaces are most commonly found over oceans or when abundant low level moisture is present elsewhere; and in the synoptic location shown (on the downstream side of a ridge aloft and extending to the next trough).

Over oceans, the clouds (at "D") on the north side of the interface are convective cells with little or no persistence of debris at their tops. The clouds at "C" are similar convective cells with "outflow layer" debris spreading and persisting in a stable layer near the cloud top level. At middle levels in the jet stream baroclinic zone along "E" there is a strong temperature gradient with cooler air over the "D" type clouds, and warmer air over the "C" type clouds. Thus, there is a stable layer over the "C" area clouds - perhaps even an inversion.

I have found that the cloud interfaces - such as "E" - are usually parallel to the associated jet stream axes, but are usually farther south than the jet axis (typically from 1 to 3 degrees of latitude). The position of the interfaces that I studied carefully, corresponded best with the axis of maximum winds at 700mb.

My observations are that the jet stream related lower cloud interfaces are most likely to be present when the associated jet stream is: (1) well defined, (2) in a channeled or nearly channeled configuration, (3) vertically deep in the atmosphere, and (4) over the ocean or some other source of abundant moisture.

Over land, similar boundaries - as that located at "E" - sometimes occur; but the cloud patterns are often more complicated and depend upon variations in available moisture. There are differences in cloud type and large diurnal variations superimposed upon the synoptic pattern changes. Many times - without sufficient moisture - there are no low or middle clouds present, although the jet stream conditions are adequate. When low level moisture is sufficient, the most prevalent cloud condition is to have an overcast deck of stratocumulus under the cyclonic side of the jet axis (in position of the "D" clouds in Fig14). When even more moisture is present, there will be a deck of altostratus (or low level convective clouds with a layer of altostratus near the tops) on the anticyclonic side of the jet axis, in addition to the overcast stratocumulus. When this occurs, the altostratus will have a well defined boundary near the axis of maximum winds at its level, and a clear band separating it from the stratocumulus.

At times - when the jet stream upstream was **initially** channeled with a low cloud interface present - but is changing to advection type (NVA), there will be a period of time when the interface is still identifiable and the jet is not channeled In such cases the interface will roughly identify the jet axis position, but will not be parallel to the axis. The jet axis orientation will be rotated somewhat clockwise from that of the interface.

Such a change from channel to advection structure often occurs when the upstream ridge is building northward, (at "A" on Fig 14) or when the next short wave perturbation or speed max coming over the ridge does so at a higher latitude than the preceding one. The latter process is usually the way it happens - although the larger scale result is still that the upstream ridge has "built" northward.

Another indication of jet location and structure in the area of the "E" interface of Fig 14 is the type clouds along the frontal zone (F) that extends back from cloud system "B". Along that zone, cirrus topped clouds - likely convective in nature - extend down the frontal zone as far as "V". Note that north of "V" the jet axis is of the advection type - PVA (the arrows shift to the left as they progress downstream toward "B".) Upstream from "V" (to the west) the jet is channeled as far as point "R". Even without the occurrence of the low cloud interface at "E", we could guess that the jet was well defined and channeled upstream from the frontal zone at "V". If the jet were weaker and less well defined along "E", and strong and channeled on the front side of the trough (from "V" to "B"), convection or persistent high clouds would not likely extend down the frontal zone to "V". This relationship between the frontal zone weather and the upstream jet is a good one, EXCEPT over land during the spring and summer seasons during the afternoon and evening hours.

Rule 3 accounts for about 25% more cases of jet stream location.

4. | Rule 4 is to interpolate the jet axis position between places it could be located using rules 1 thru 3.

This doesn't mean draw a straight line. We must take into account likely curvature of the axis based upon the long wave synoptic environment. We must remember to shift the axis polward where PVA is indicated, and equatorward where NVA is indicated.

Using rule 4 can enable us to locate the jet axis another 10% of the time.

Applying rules 1 thru 4 will allow us to locate the axes of the jet streams with typical synoptic scale accuracy from 75% to 85% of the time.

Of the remaining 15% to 25% of the time, when the jet stream axis cannot be located with synoptic scale accuracy; part of the cases will be missed because the relationships and rules presented here are not good enough. One of the following situations is likely:

a. There is no "jet axis" over the region concerned. This can occur two ways:

    (1) There is a wide zone of strong winds, but the maximum wind speeds are not concentrated in a narrow enough zone to be identified as a "jet axis".

    (2) The jet stream and its associated baroclinic zone isn't continuous between two well defined segments. Often, over an upper air ridge, the air spreads with difluence and slows into a region with no well defined baroclinity (and no strong winds).

b. There is not sufficient moisture present to produce clouds, even though the jet stream is strong and well defined.

c. High level moisture is present, with a well defined moist-to-dry border along a channeled jet axis, but clouds have not formed.

    In such cases, cirrus will often form suddenly on the right side of the jet axis, as the region of strong winds and moist/dry boundary progresses downstream over a mountain range. A high - cold - deck of cirrus, which we refer to as "lee-of-the-mountain-cirrus" will form downstream from the mountain range, and to the right of the jet axis. Such cirrus decks are <u>not always</u> on the right side of the jet axis; the occurrence seems highly dependent upon where the high level moisture is located within the strong wind zone. Lower level "lenticular" mountain wave clouds may form under such a cirrus deck. When the high level cirrus is confined to the right side of the jet axis, the lower level lenticular wave clouds will often extend out from under the cirrus well to the left side of the axis.

    Cirrus formations similar in appearance & behavior to the "lee-of-the-mountain-cirrus" decks, will also form suddenly as such a jet stream moisture deck passes over a lower level trough or frontal zone.

Airflow Through Midlatitude Cyclones and the Comma Cloud Pattern

TOBY N. CARLSON

Department of Meteorology, The Pennsylvania State University, University Park 16802

(Manuscript received 30 May 1979, in final form 17 April 1980)

ABSTRACT

Airflow through a developing midlatitude disturbance is analyzed in a relative-wind isentropic system in order to provide insight into how the cloud pattern evolves into the familiar comma shape. The model presented makes use of various concepts such as that of the conveyor belt and explores the relationship between the configuration of the major airstreams and such features as the jet streams and the dry tongue. The model also relates vertical motion and precipitation to the origin and vertical displacement of the airstreams and attaches special significance to airstream boundaries, which manifest themselves as sharp discontinuities in cloud and weather patterns.

1. Introduction

Present-day meteorological satellites in combination with conventional radiosonde observations afford a fresh opportunity for studying the movement of airstreams through wave cyclones. Remarkably, satellite imagery reveals much of what early Scandinavian meteorologists, such as Bjerknes and Solberg (1922), were able to describe without recourse to satellite imagery or even to upper air data.

To summarize the cloud distribution represented in early cyclone models, the initial stages of wave cyclone development are accompanied by the formation of a crescent-shaped band of dense layer cloud situated largely on the cold side of the surface fronts but within ascending air whose origin lies at low levels on the warm side of the fronts. Typically, a broad area of cloud cover extends north of the warm front and a narrower zone of cloud trails along the cold front, as illustrated by the solid outline in Fig. 1. This cloud region is characterized by stratiform types which are produced by condensation within layers of relatively warm, moist air which originate within the lower troposphere south and east of the surface low and which undergo large-scale slant ascent over the colder air. Precipitation is widespread north of the warm front but the outer edges of the cloudy region north of the warm front consist only of thin layers of upper and middle cloud.

With the advent of satellites, numerous observational studies have added to but not substantially modified the earlier concepts. Recent satellite evidence indicates that the development of a disturbance is often accompanied by an expansion of the cloud shield, most notably toward the west and southwest of the surface low center (the dashed area labeled in Fig. 1). The cloud distribution may come to resemble the shape of a comma, the so-called comma cloud, in which the head is represented by the northern and western part of the cloud shield (the areas labeled 1 and 2 in Fig. 1). In later phases of cyclone development, shallow stratiform cloud extends to the west and southwest of the middle and upper cloud (the area labeled 0 in Fig. 1), while a cloud-free zone (labeled 3 in Fig. 1) intrudes into the comma head region, cutting off the westward cloud protrusion from the main body of the cloud shield. Ultimately, the occlusion stage is reached and these easily describable features become lost in a large swirl of cloud surrounding the low center.

2. Analysis procedure

Let us now consider the motion through a typical wave disturbance. In order to follow the trajectories of the air, the motions will be analyzed in a *relative-wind isentropic system* in which the winds are represented on surfaces of constant *potential temperature* (θ) in dry regions and on surfaces of constant *wet-bulb potential temperature* (θ_w) where large-scale slant ascent is occurring in saturated regions. To avoid employing a succession of charts at different times, the winds are represented in the system *relative* to the movement of the large-scale disturbance. The assumption that the system translates steadily without change of speed or shape enables one to subtract vectorially the system's phase speed from the measured winds and to utilize these relative winds to construct a set of streamlines which are also trajectories relative to the movement of the

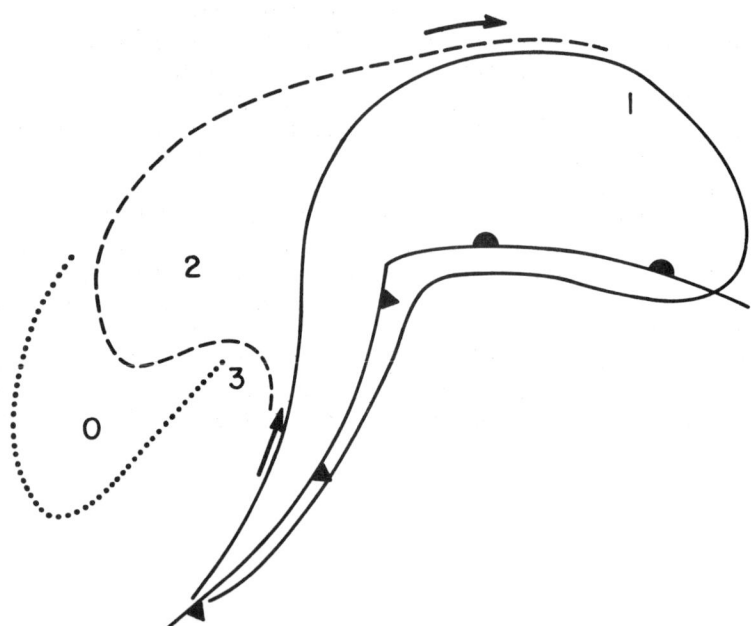

FIG. 1. Schematic illustration of cloud shield associated with a midlatitude wave disturbance. Early stage of cyclone development is associated with area of dense middle and upper layer cloud contained within solid boundary and labeled 1. Dashed line encloses area labeled 2 which corresponds to westward extension of cloud canopy. Area enclosed by dotted border labeled 0 denotes layer of thin status within the boundary layer. The dry tongue is represented by the narrow zone labeled 3. Arrows indicate location of major upper-tropospheric wind maxima.

system.[1] Relative isentropic analysis was first used by Eliassen and Kleinschmidt (1957) and later by Green et al. (1966), Carlson and Ludlam (1968), Browning and his associates (Browning and Harold, 1969; Browning, 1971; Harrold, 1973; and most recently by Atkinson and Smithson, 1978) to study the motion of airstreams through wave cyclones. Reiter and his associates (Reiter and Mahlman, 1965; Reiter et al., 1965) and Danielsen (Danielsen and Bleck, 1967; Danielsen, 1974) have also performed extensive analyses in isentropic coordinates, for the most part using a succession of maps, rather than the phase speed correction, to determine the three-dimensional trajectories of air motion through wave cyclones. By avoiding the use of 12 h time intervals for computing trajectories, the relative-wind assumption is suitable for representing boundaries between airstreams of differing origin.

Relative isentropic analysis will be used in this paper to describe rather generally the motions of air through a typical intensifying wave cyclone, with emphasis on showing how the large-scale layer cloud relates to these motions. Several cases of midlatitude wave cyclones have been examined by the author using isentropic analyses in conjunction with classroom exercises in synoptic meteorology courses at the Pennsylvania State University. This paper represents a conceptual synthesis of these analyses and was provoked by a dilemma, which was to show how the cloud extension west of the cyclone could have formed when there was evidently no air reaching that area from the warm sector and when the air to the west of the comma head had been sinking and was essentially dry.

It is the purpose of this paper to summarize the findings from these analyses using one case study as an illustration. For clarity and simplicity, only a limited analysis is presented. Also, plotted data will be omitted and certain fields will be represented schematically (e.g., wind speed maxima are located by arrows).

3. A case study

a. General situation

The situation used to illustrate the evolution of a comma cloud occurred on 5 December 1977. In Fig. 2 a conventional isobaric surface map is presented showing a mature wave cyclone which had been intensifying rapidly during the previous 12 h. At 1200 GMT snow was beginning to fall over Pennsylvania (P). During the next 24 h moderate to

---

[1] This is mathematically equivalent to setting the local derivative of potential temperature $\partial\theta/\partial t = -C_x(\partial\theta/\partial x)$, where C_x is the eastward phase speed of the longwave system.

heavy snow accumulations were recorded over a region extending from northern Pennsylvania to New England (N). In this case the comma head corresponded to the cloud shield north and west of the surface low center. The tail of the comma extended a short distance toward the south.

The scalloped area in Fig. 2 conforms to the outline of the cloud pattern in the infrared satellite enhancement of Fig. 3. In subsequent discussion, reference will be made to various features associated with the cloud pattern which are labeled numerically or by letters on the satellite picture. Emphasis will be directed toward an analysis of the airflow through the main cloud shield in the figure. Features not associated with large-scale slant ascent within the major cloud shield will be ignored. They include the low-level stratus (0); the narrow strip of cloud located to the west of the main cloud shield (4); the squall line to the south of the warm front (5); and the wide band of cirriform cloud associated with the subtropical jet (6).

It can be seen from Figs. 2 and 3 that the low center (L) was situated at the north end of the dry tongue (3), the latter forming an almost cloud-free indentation into the base of the comma head. One can observe a slight fissure in the cloud along a line northeast from point 3 in Fig. 3. Highest cloud tops, represented by the dark enhancement on the satellite photograph, were present over western Pennsylvania (between P and 1). Although the comma

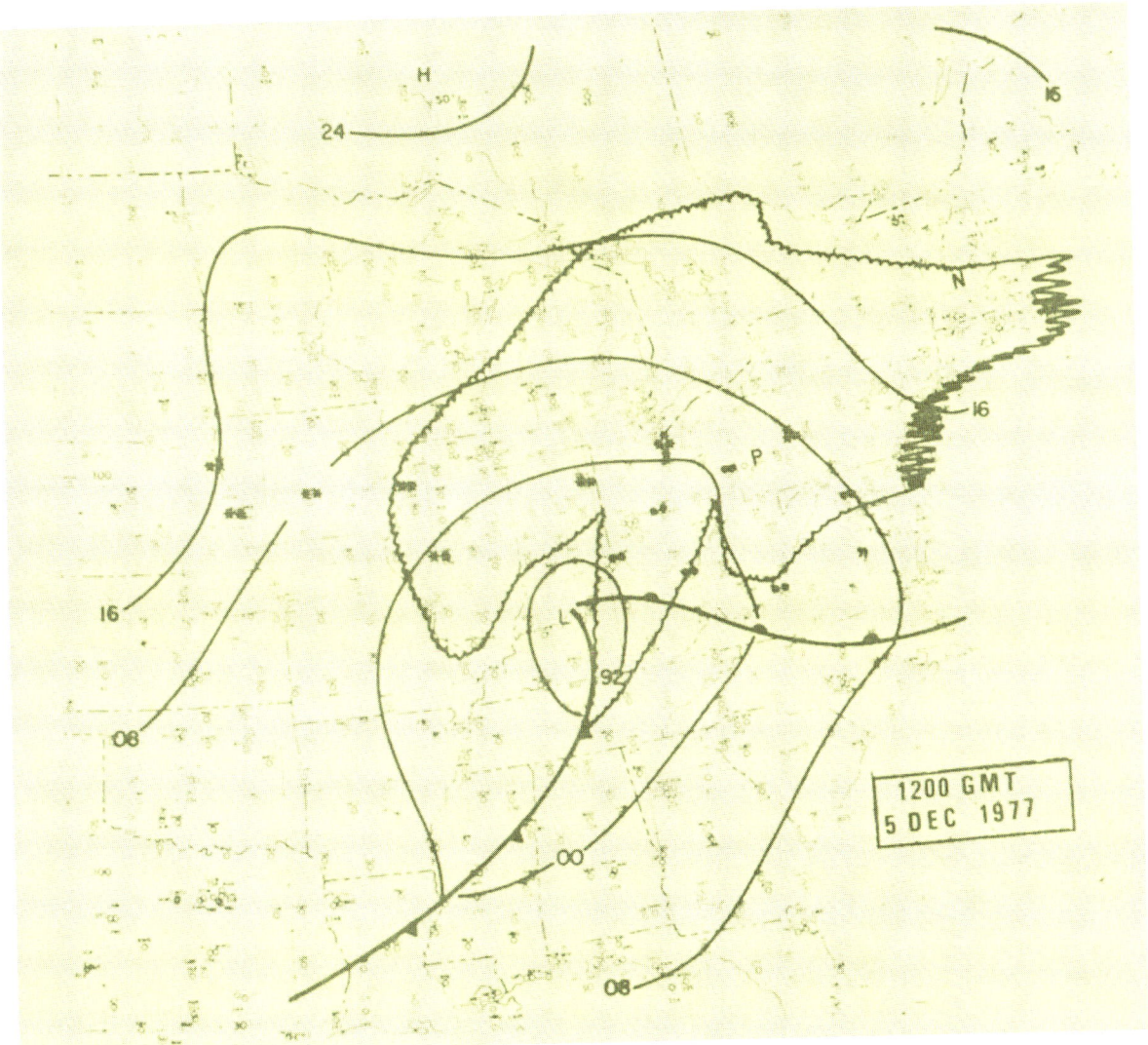

FIG. 2. Surface isobaric chart (labeled conventionally in mb) for 5 December 1977. Scalloped region encloses region of dense middle and high cloud, the major cloud shield corresponding approximately to that in Fig. 3. Symbols represent conventional meteorological designations. The letters P and N refer to place locations (Pennsylvania and New England) mentioned in the text.

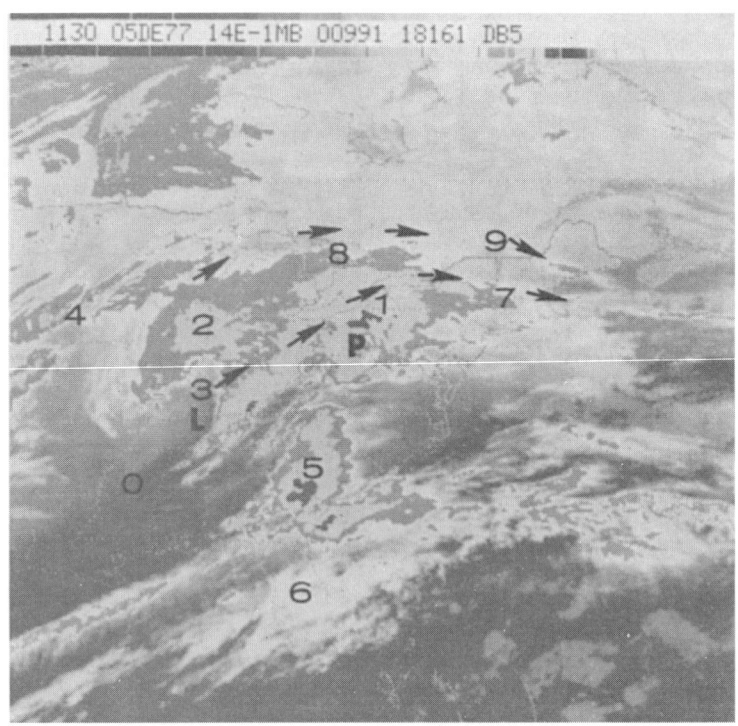

FIG. 3. Enhanced infrared GOES satellite image for 1130 GMT 5 December 1977. Arrows represent axes of major jet streams and correspond closely to portion of limiting streamlines of Figs. 4 and 5. Various letters and numbers indicate features referred to in text.

tail was abbreviated, the comma head exhibits the westward protrusion in the area of point 2. The southern part of the cold front was almost free of cloud but widespread precipitation was occurring north of the warm front (in the general area of P).

b. *The warm conveyor belt*

On the 305 K isentropic surface (Fig. 4), warm, moist air entered the warm sector from the south and flowed north of the warm front. After reaching saturation, the air ascended to the upper troposphere following a saturated isentropic surface with a wet-bulb potential temperature of $\theta_w = 15°C$. The flow remained on the right (east) side of the southwesterly jet, which is represented on the figure by a streamline that coincides with the stream of arrows between points 3 and 7 in Fig. 3. The saturated airstream reached high levels over New England near the ridge and descended toward the downstream trough. Browning and his associates in England (see Browning, 1971) attach special significance to the moist, ascending branch of the flow, referring to it as the *warm conveyor belt*. In subsequent discussions we will refer to the warm conveyor belt as that air which originated far south of the low in the warm sector, ascended toward the north, achieved saturation near or north of the warm front, where it rose more rapidly, and joined the upper-level westerly flow northeast of the low center.

Although it is difficult to assign exact boundaries to the warm conveyor belt, the latter corresponds in Fig. 4 to a 200–300 mb deep layer, which was situated to the right (east) of the streamline labeled LSW. Because the warm conveyor belt was derived from air that originated within the convective mixing layer, the isentropic surface was chosen to lie somewhere near the top of that layer so as to minimize the non-adiabatic effects occurring near the surface. However, it is likely that the deep convection associated with the vigorous squall line was causing the flow of moist air to be modified south of the warm front.

The flow within the warm sector in Fig. 4 originated largely within a relative easterly flow at low latitudes which turned northward ahead of the disturbance. The left (west) edge of the warm conveyor belt, coinciding with that streamline labeled LSW, originated farthest south and most nearly approached the jet in the vicinity of the disturbance. This streamline constituted a western *limit* to the meridional exchange on that potential temperature surface. Therefore, it will be called the *limiting streamline* for the *warm* conveyor belt.

Because of its southernmost origin, air immediately to the right of the limiting streamline for the warm conveyor belt possessed the highest moisture content in the southerly flow of moist air. The air along this part of the flow also experienced the greatest vertical displacement, in accordance with the fact that the isentropes sloped upward toward the colder air to the west. Thus, air just east of the limiting streamline for the warm conveyor belt experienced the strongest vertical motion

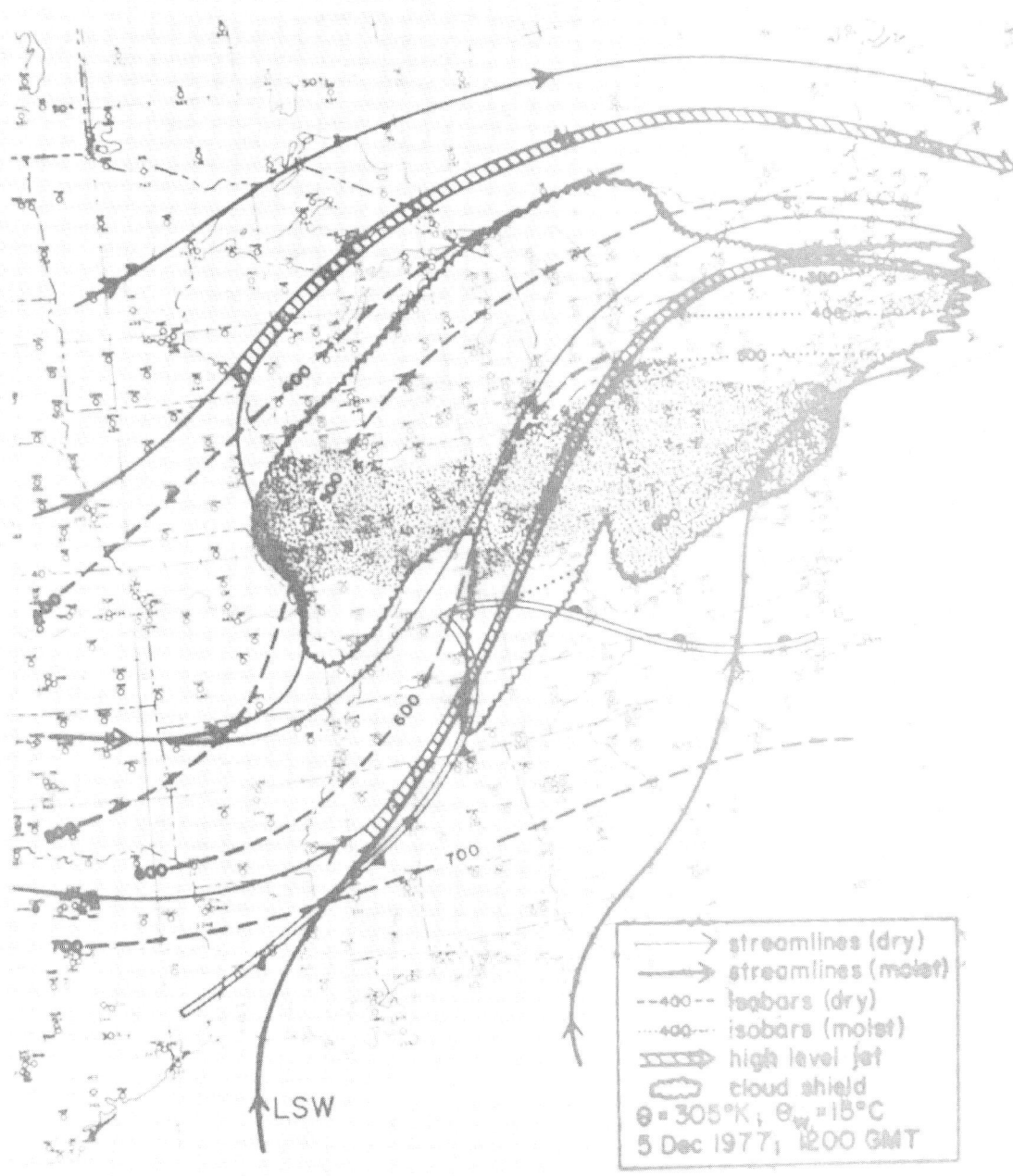

FIG. 4. Relative wind isentropic analysis for 1200 GMT 5 December 1977. Streamlines pertain to relative wind velocity in which the phase speed of the long wave has been subtracted out (13 kt eastward). Solid streamlines represent flow on the 305 K isentropic surface and dashed lines are isobars on the θ surface (labeled in mb). The heavy streamlines refer to moist flow reaching saturation on the $\theta_w = 15°C$ wet-bulb potential temperature surface and thin solid streamlines represent air which does not reach saturation. The limiting streamline for the warm conveyor belt, labeled LSW, defines the western edge of a flow which is achieving general saturation within the scalloped border of dense middle and high cloud to the east of the jet. Stippling represents sustained precipitation associated with the cloud shield. Double-shafted hatched streamlines indicate location of major upper-tropospheric jet axes.

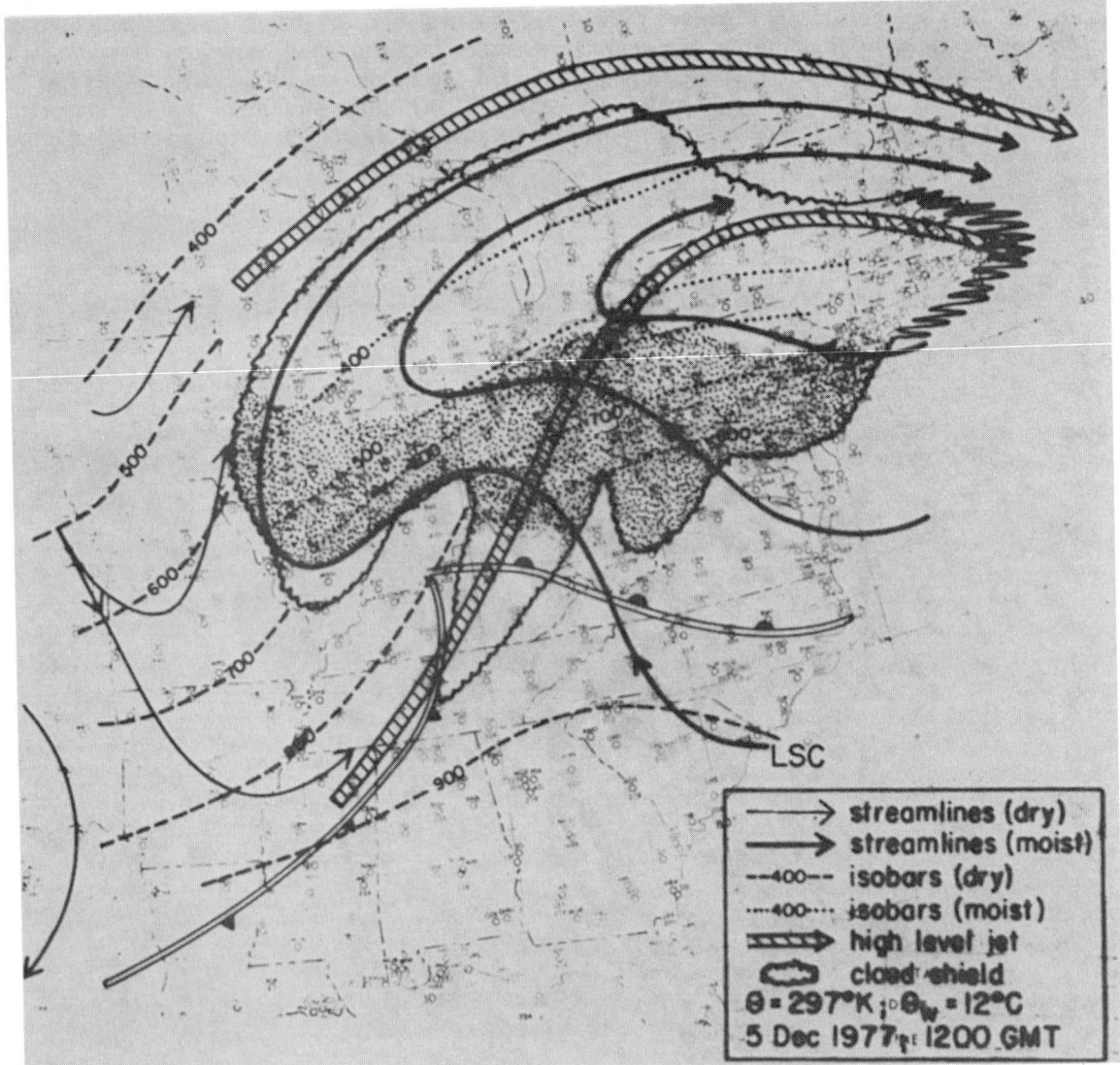

Fig. 5. As in Fig. 4 except for the $\theta = 297$ K dry isentropic and $\theta_w = 12°C$ moist adiabatic surfaces. Saturation is occurring within the cold conveyor belt and to the east of the limiting streamline for the cold conveyor belt, labeled LSC.

because the largest displacement occurred in conjunction with the higher wind speeds near the jet. Successive streamlines to the east were less moist, rose more slowly and through a smaller vertical distance. The right edge of the conveyor belt represented air which just barely achieved saturation. Farther to the east no cloud was formed because the air was too dry and the ascent was insufficient to produce condensation.

Immediately to the left of the limiting streamline, the air was very dry, having descended from higher levels west of the trough axis to the area of the dry tongue to the west. Air flowing over the region of the dry tongue ascended over the top of the cloud shield west of the limiting streamline which therefore constituted a major axis of confluence near the cold front along which air of highly differing origins joined together. This confluence line also corresponded closely to the axis of the polar front jet whose core of maximum winds lay within the cloud-free zone west of the front.

c. *The cold conveyor belt*

Consider now a lower isentropic surface such as $\theta = 297$ K (Fig. 5). The surface did not exist very extensively in the warm sector where the potential temperature was generally above 297 K in the surface layers. North of the warm front, however, a substantial depth of air was present with a potential temperature below $\theta = 297$ K. Actual winds in that layer were either weak southerly or easterly. Since the relative winds were generally from the east, much of the air on the cold side of the warm front flowed toward the region north of the low center and beneath the warm conveyor belt. The stream of air, which will be called the *cold conveyor belt*, originated in the descending air associated with the previous high pressure system.

Air following the cold conveyor belt flowed westward beneath the warm conveyor belt north of the surface low (Fig. 5). Saturation in the cold conveyor belt was aided by precipitation falling from the warm conveyor belt and by subsequent ascent of air within the former near the western side of the latter. As in the warm conveyor belt, ascent in the cold belt was most pronounced near the location of the southwesterly jet streak where the isentropic surfaces sloped upward most steeply toward the cold air.

Air ascended within the cold conveyor belt in a manner similar to that within the warm belt. On the moist isentropic surface ($\theta_w = 12°C$), the former ascended rapidly and turned sharply northward, and folded beneath the latter at high levels near the ridge. The limiting streamline for the cold airstream, as drawn in Fig. 5 and indicated in Fig. 3 by the stream of arrows crossing the Great Lakes between points 8 and 9, extended to the north and west of the warm conveyor belt along the northern edge of the cloud shield. Although the data are insufficient to determine the exact location of a limiting streamline for the cold airstream (the left-hand edge of this airstream), the northern edge of the cloud shield was sharply defined in the ridge by the confluence be-

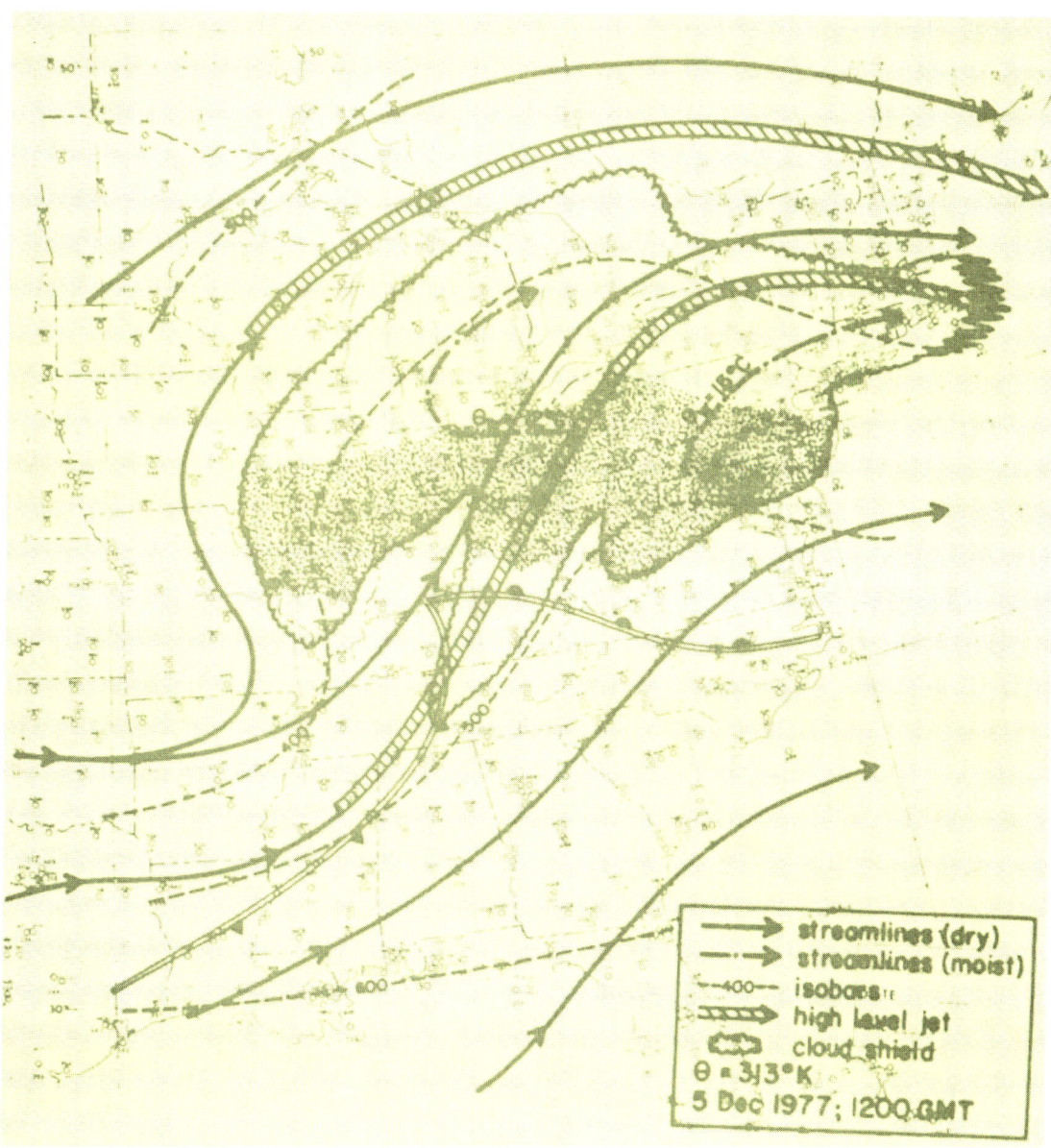

FIG. 6. As in Fig. 4 except for $\theta = 313$ K. Saturation does not occur following the solid streamlines, but the dot-dashed streamlines represent saturated air rising from lower isentropic surfaces through the 313 K surface, following saturated adiabatic paths along the warm ($\theta_w = 15°C$) and cold ($\theta_w = 12°C$) conveyor belts.

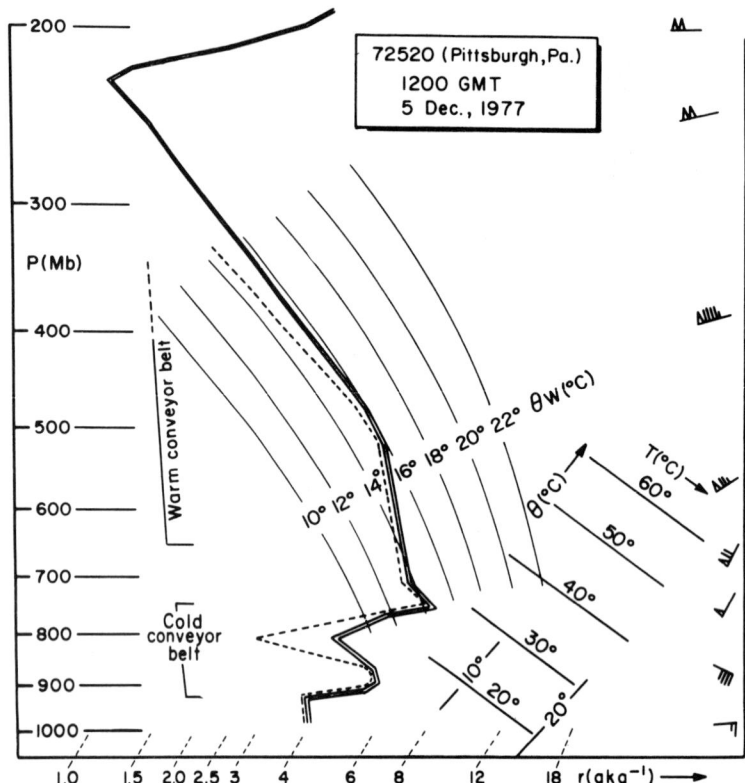

FIG. 7. Skew-T Log$P$ temperature (double solid lines) and dewpoint (dashed line) for Pittsburgh 1200 GMT 5 December 1977. Wind speed and direction plotted at right in conventional notation.

tween the moist airstreams and a drier one to the north. Located along the confluent zone and in the ridge was a westerly jet maximum (Fig. 5).

d. The dry stream

At higher θ surfaces than $\theta = 313$ K (Fig. 6), the air reaching the cloud shield from the west originated not in the lower troposphere but in the upper troposphere west of the trough. Consequently, the air was quite dry, especially along the path immediately to the west of the warm conveyor belt because of strong prior subsidence. The streamlines near the cloud shield were strongly diffluent as the dry air tended to flow around regions of strongly rising saturated air. Thus, air reaching the cloud shield on the 313 K surface did not achieve saturation. Condensation was nevertheless occurring on that isentropic surface within the cloud shield as the result of the rapidly ascending conveyor belt flow. The moist air reaching the isentrope can be viewed as having originated at lower isentropic surfaces but was flowing along a path to higher isentropic surfaces along surfaces of constant θ_w.[2] Diffluence

of dry air in the vicinity of the cloud shield, observed on both the $\theta = 305$ and 313 K surfaces (Fig. 5 and 6), is therefore thought to reflect the spreading of dry flow lines to accommodate this rising stream of saturated air from lower θ surfaces.

Both moist conveyor belts can be identified on the 1200 GMT Pittsburgh sounding (Fig. 7—located at P in Fig. 3). The warm conveyor belt existed primarily above 750 mb, although it is uncertain where the upper boundary should be drawn because there is no clear indication of a top to the moist layer. [As pointed out by Browning (1971) a decrease with height in θ_w can lead to significant mixing of the unsaturated air aloft with the moist air below.] Below 750 mb, air within the cold conveyor belt with a relatively low θ_w was flowing westward at approximately right angles to the flow aloft. A shallow unsaturated segment existed between these airstreams, presumably because the combination of ascent in tne lower airstream and evaporation of precipitation falling from above was insufficient to saturate air throughout the entire depth of the lower layer.

Air in the cold conveyor belt flowed westward beneath the warm conveyor belt and beneath the dry airstream which had descended from west of the upper trough. Now, air immediately to the west of the limiting streamline for the warm conveyor

---

[2] The presumption that no mixing was taking place is certainly questionable. There was undoubtedly some degradation in the integrity of the flow, so that a considerable subjective judgement must be exercised in maintaining a continuity of flow.

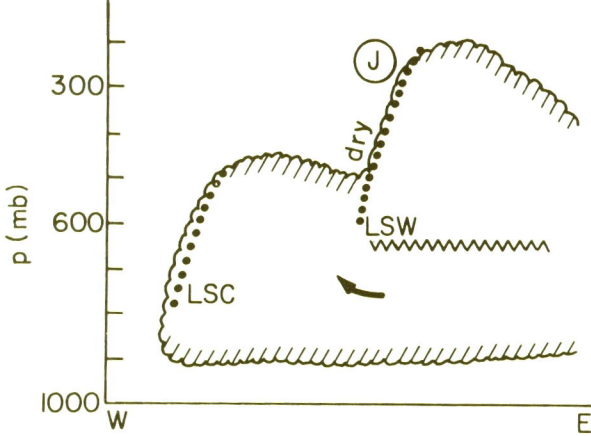

FIG. 8. Schematic cloud cross section west (W) to east (E) intersecting the limiting streamlines (dotted lines labeled LSW or LSC) for the warm and cold conveyor belts, respectively. Scalloping outlines the region of dense layer cloud. The arrow suggests the vertical and horizontal movement of the cold conveyor belt under a tube of dry air beneath the jet axis (labeled J), into the western part of the cloud extension (comma cloud shield). The warm conveyor belt is located to the right of this jet and above the serrated border.

belt coincided with the southwesterly jet whose core of maximum winds was located over the cloud-free intrusion. A tube of very dry air was situated immediately below the jet streak and just to the west of the elevated warm conveyor belt cloud (Fig. 8).

4. A conceptual model

The technique of relative-flow isentropic analysis, which reveals the flow of air on isentropic surfaces relative to the movement of a mature midlatitude cyclone, provides insight into how the cloud pattern as seen in visual and infrared satellite imagery evolves into the familiar comma shape. A typical midlatitude system, portrayed in Figs. 9 and 10, contains three major airstreams of widely differing origins. Two of these originate at low levels and the third comes from the upper troposphere west of the trough axis. The first airstream, referred to as the warm conveyor belt, originates in relative easterly flow at low latitudes. This air turns northward to flow approximately parallel with the cold front and ascends while turning anticyclonically above the warm front, joining the upper-level flow northeast of the surface low near the upper-level ridge. The clouds associated with the southwesterly flow along the cold front and north of the warm front form the comma tail and body. The second airstream, referred to as the cold conveyor belt, originates in the anticyclonic low-level flow to the rear of a surface high east of the surface low. This air approaches the low from the east and rises rapidly as it moves toward the west from underneath the warm conveyor belt. It continues to rise and turn anticyclonically toward the northeast and merges with (or folds beneath) the warm conveyor belt in the upper troposphere near the ridge. The low- and middle-level clouds associated with the portion of the cold conveyor belt situated west and north of the surface low form the westward extension of the comma head.

The distinctly sharp edge of the cloud pattern on the west edge of the comma tail, as well as the northern and northwestern portion of the comma head comprising the anticyclonically curved cirrus shield northeast of the low, is due to the confluence of the moist air with a third major airstream. This third airstream originates in high tropospheric or lower stratospheric levels downstream from the ridge far to the northwest of the surface low. A portion of the airstream descends anticyclonically to the lower troposphere west of the trough and east of the surface high pressure system (Fig. 10). Some of this air splits with the anticyclonic stream, crosses the trough axis, flows parallel to the warm conveyor belt within the cloud-free zone, and ascends above and also around the middle-upper cloud deck of the cold conveyor belt. A southwesterly jet streak whose core lies along the eastern edge of the dry tongue extends along the line of confluence between the warm conveyor belt and the dry airstream. Southerly air flowing over the warm sector at high levels above the warm conveyor belt originates at middle and upper levels within the tropical easterlies east of the upper trough axis. This air may be unsaturated when it reaches the warm sector, so that it must flow over or around the conveyor belt cloud shield. In some instances, such as when the southerly flow above the moist part of the conveyor belt originates in a region of a convective disturbance or is modified by cumulus convection along the way, condensation may occur in the form of an extensive cirrus shield over the warm sector or within a narrow cloud band along the comma tail, as indicated by the streaky lines in Fig. 9. Since this upper-tropospheric flow also experiences a strong confluence with the dry flow to the west, the western edge of the upper cirrus shield in the warm sector will be extremely sharp but the air also tends to flow around the cloud shield in a diffluent pattern.

The top of the cold conveyor belt cloud deck is generally at a lower elevation than the warm conveyor cloud band, as shown in Fig. 8. The western border of the latter cloud is identifiable on satellite photographs, particularly at low sun angles when the terraced effect in the cloud levels is brought out in relief by shadows.[3] Air within the former cloud

---

[3] This author has observed these features on numerous occasions. In most cases the main jet streak east of the trough axis corresponds closely to the raised wall of cloud that lies along the western edge of the warm conveyor belt.

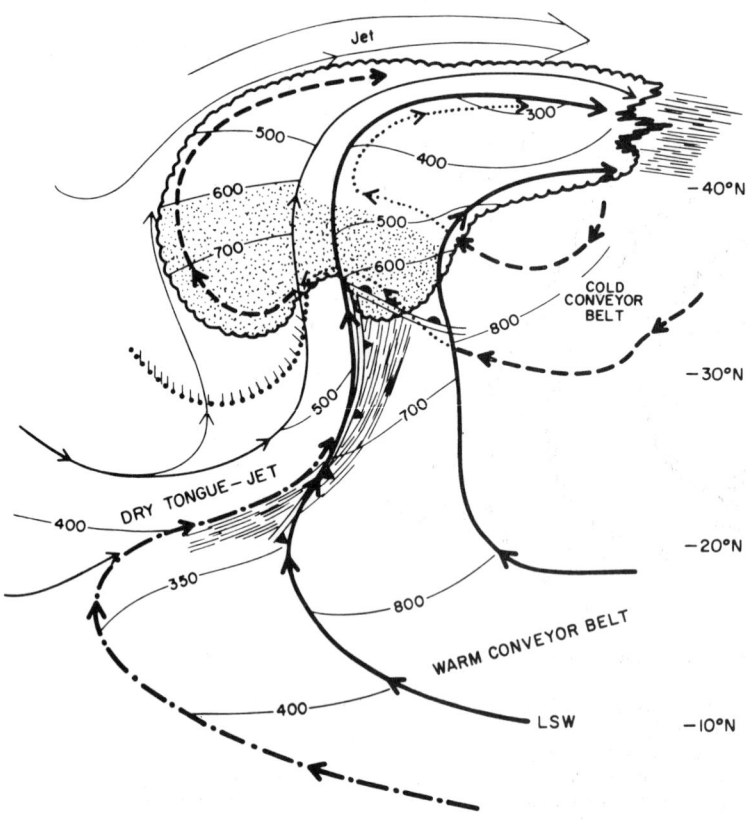

AIRFLOW THROUGH MID-LATITUDE WAVE CYCLONE

FIG. 9. Schematic composite of airflow through middle-latitude cyclone. Heavy solid streamlines depict airflow at top of the warm conveyor belt. Dashed flow represents cold conveyor belt (drawn dotted when it lies beneath the warm conveyor belt or dry airstream.) Dot-dashed flow represents air originating at middle levels in tropics. Thin solid streamlines pertain to dry air which originates at upper levels west of the trough. Thin solid lines denote the heights of the airstreams (mb) and are approximately normal to the direction of the respective air motion; (isobars are omitted for the cold conveyor belt where it lies beneath warm conveyor belt or beneath the jet stream flow). The region of dense upper- and middle-level layer cloud is represented by scalloping and sustained precipitation by stippling. Streaks denote thin cirrus. The edge of the low-level stratus is shown by the curved border of small dots with tails. The major upper tropospheric jet streams are labeled JET. The limiting streamline for the warm conveyor belt is labeled LSW. See text for further explanation.

layer flows westward almost perpendicularly to the direction of the warm conveyor belt in the direction of the cold air to the west of the low. Thus, the inception of the westward cloud extension is also associated with rapidly veering winds with height and therefore with strong warm air advection north and west of the surface low. With the extension of the cloud shield to the west, a substantial amount of precipitation is recycled from the warm air and deposited in the presence of relatively cold air west of the low.

5. Relationship to other models

The English School has devoted a considerable amount of discussion to the properties of the warm conveyor belt, treating the airstream as a relatively discrete entity. Browning (1971) alludes to a cold conveyor belt, a descending easterly airstream which flows north of the warm front and is responsible for a redistribution of moisture (derived from evaporated precipitation falling from the warm conveyor belt) within the system. Reiter *et al.* (1965; see Fig. 36) present a model possessing some aspects discussed in this paper, in which a warm conveyor belt is shown to be confluent with a flow of rapidly moving air whose origin was in the high troposphere west of the trough axis. Reiter *et al.* (1965) also indicate that a branch of this upper-tropospheric flow descends from south of the jet into the low troposphere behind the surface cold front (Reiter and Mahlman, 1965; Fig. 7). Eliassen and Klein-

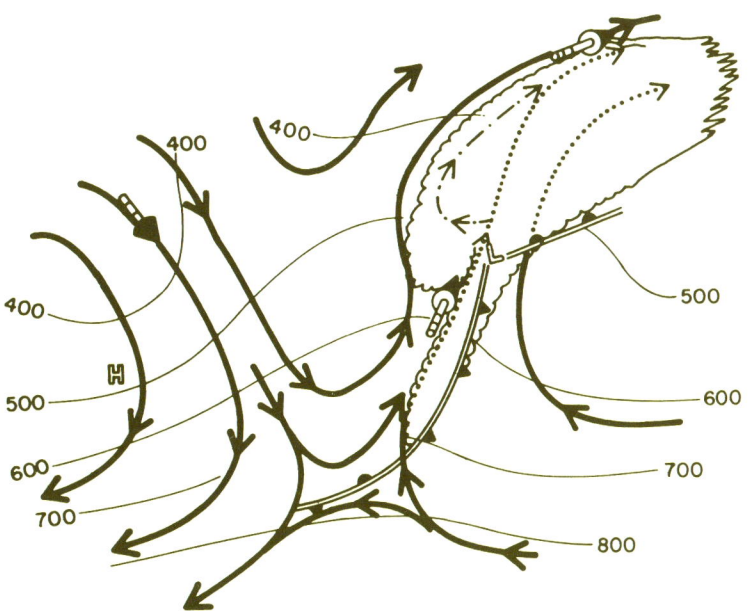

Fig. 10. Schematic illustration of relative flow on a dry isentropic surface. Thin solid lines are isobars; these are discontinued in regions of saturation where the flow is drawn dotted for the warm conveyor belt, and dot-dashed for the cold conveyor belt. The dry airstream originates at high levels in the northwest and descends toward the trough axis. There it splits into two branches, one branch descending into low levels west of the trough and the other flowing around the trough to ascend in a narrow stream over the western extension of the cloud shield (scalloped border). The symbols L and H, respectively, refer to the locations of the surface high and low pressure centers. Fronts are depicted in conventional symbols. Jet maxima are represented by solid arrow segments.

schmidt (1957) were perhaps the first to depict this airstream on isentropic surfaces along which the trajectories (Fig. 35) form an "upslide" cloud north of the warm front. Their analyses show a rather sharp transition between the warm conveyor belt and a cold, ascending airstream north of the warm front. This cold air flows westward beneath the warm conveyor belt forming the westward extension of the upslide cloud shield west of the surface low. However, the rising cold air is shown to turn cyclonically toward the south and begin sinking behind the cold front along the southern edge of the comma head. Danielsen (1974) has discussed the role of the descending dry airstream, which is associated with an upper-tropospheric jet and which originates near the base of the stratosphere. He shows that this air arrives at low levels to the east of the trough axis where it joins a moist airstream from the south. Danielsen's analyses do not seem to show a clearly defined cold conveyor belt and, with the exception of a warm conveyor belt type of flow, the airstreams he depicts do not flow unobstructed through the disturbance but, instead, impinge on one another along sharp boundaries. Palmén and Newton (1969; Fig. 10.20) present an overview of both the dry and moist streams which appear as two symmetric branches of a meridional air exchange between the upper troposphere at high latitudes and the lower troposphere at low latitudes. Both branches of the air exchange are anticyclonic. Recently, Grotjahn (1979) was able to simulate in an analytic model the conveyor belt and dry airstream in a wave disturbance. His Fig. 24 is remarkably similar to the Palmén and Newton (1969) figure and also closely resembles the model presented here. Grotjahn's flow pattern gives some indication of a cold conveyor belt, but the trajectories do not indicate whether that air turns southward and descends behind the cold front or turns anticyclonically toward the north and ascends to the high troposphere, in the ridge, as suggested in this paper. Thus, the motion of the cold conveyor belt to the west of the surface low represents the main difference between the model presented here and those previously described in the literature. Our analyses consistently indicate that if the cold air north of the warm front is warmer than the advancing cold air from the west, the former will rise rapidly as it moves west of the surface low. Unless the system has entered the final occlusion stage, the air aloft will have a substantial relative southerly and westerly component at high levels which will enable at least a part of that flow to circulate toward the north and east after it reaches middle levels.

6. Implications for cyclone development

In this paper, the frontal zone east of the upper trough is associated with a confluence of discrete airstreams that differ greatly in their properties and in their geographical origins. Consequently, there is a sharp boundary across the confluence line which manifests itself as a discontinuity in cloud and moisture. Eliassen (1962), among many others, has shown that confluent deformation of the isotherms leads to a rapid formation of temperature fronts. Others (Danielsen, 1968; Hoskins, 1971; Shapiro, 1978) show that confluent deformation in the high troposphere can lead to the formation of an upper-level front and a fold in the tropopause near the jet just downstream from the ridge. In this region strong subsidence at the base of the stratosphere causes portions of the lower stratosphere to be extruded into the troposphere within the upper-tropospheric front.

The present model indicates that strong confluence between the dry and moist airstreams occurs in the lower and middle troposphere along the location of the surface cold front and in the upper troposphere near the ridge along the northern border of the cloud shield north and northeast of the surface low center. Shapiro (1978) presents evidence that upper-tropospheric frontogenesis begins in the upstream ridge in association with strong sinking motion between the ridge and downstream trough. Hoskins (1971) shows that intrusions of upper tropospheric air into the middle troposphere and the formation of an upper level front occur in response to confluent deformation at high levels. In the present case, that confluence is between the conveyor belts and the dry airstream. Simmons and Hoskins (1979) show that cyclone development can propagate a succession of unstable baroclinic waves upstream and downstream from the site of initial development. One might, therefore, imagine that upper-tropospheric frontal development, which occurs in the ridge, leads to development ahead of the downstream trough. The interplay of confluent moist and dry airstreams would promote further upper-tropospheric frontogenesis in the downstream ridge and subsequent cyclone development in the next downstream trough.

7. Conclusion

The conceptual model discussed provides a physical picture of the movement of air through midlatitude systems. Air flow is represented by the motion within discrete airstreams. In the model, frontal zones arise from the confluence of airstreams with differing properties and from vastly different origins. In the lower troposphere, the cold front represents a boundary between the warm conveyor belt and the dry airstream from the upper levels west of the trough. The warm front is the boundary between the warm conveyor belt and the cold conveyor belt originating from the northeast and east. Upper-level frontal zones and jet streams are associated with the confluence between airstreams, which produce sharp boundaries in the cloud pattern along the west edge of the comma tail as well as the edge of the anticyclonically curved cirrus shield northeast of the low.

REFERENCES

Atkinson, B. W., and P. A. Smithson, 1978: Mesoscale precipitation areas in a warm frontal wave. *Mon. Wea. Rev.*, **106**, 211–222.

Bjerknes, J., and H. Solberg, 1922: Life cycle of cyclones and the polar front theory of atmospheric circulation. *Geofys. Publ.*, **3**, No. 1, 1–18.

Browning, K. A., 1971: Radar measurements of air motion near fronts. *Weather*, **26**, 320–340.

——, and T. W. Harrold, 1969: Air motion and precipitation growth in a wave depression. *Quart. J. Roy. Meteor. Soc.*, **95**, 288–309.

Carlson, T. N., and F. H. Ludlam, 1968: Conditions for the occurrence of severe local storms. *Tellus*, **20**, 203–226.

Danielsen, E. F., 1968: Stratospheric-tropospheric exchange based on radioactivity, ozone and potential vorticity. *J. Atmos. Sci.*, **25**, 502–518.

——, 1974: The relationship between severe weather, major dust storms and rapid large-scale cyclogenesis (II). *Subsynoptic Extratropical Weather Systems: Observation, Analysis and Prediction. Notes from a Colloquium: Summer 1974*, Vol. II. NCAR Rep. No. ASP-CO-3-V-2 226–241. [NTIS PB 247286].

——, and R. Bleck, 1967: Research in four-dimensional diagnosis of cyclonic storm cloud systems. Final scientific report to the Air Force Cambridge Research Laboratories, Bedford, Mass., AF 19(628)-4762, 96 pp. [NTIS AFCRL 67-0].

Eliassen, A., 1962: On the vertical circulation in frontal zones. *Geofys. Publ.*, **24**, 147–160.

——, and E. Kleinschmidt, 1957: Dynamic meteorology. *Handbuch der Physik*, Vol. 48, S. Flugge, Ed., Springer-Verlag, Berlin, 1–154.

Green, J. S. A., F. H. Ludlam and J. F. R. McIlveen, 1966: Isentropic relative-flow analysis and the parcel theory. *Quart. J. Roy. Meteor. Soc.*, **92**, 210–219.

Grotjahn, R., 1979: Cyclone development along weak thermal fronts. *J. Atmos. Sci.*, **36**, 2049–2074.

Harrold, T. W., 1973: Mechanisms influencing the distribution of precipitation within baroclinic disturbances. *Quart. J. Roy. Meteor. Soc.*, **99**, 232–251.

Hoskins, B. J., 1971: Atmospheric frontogenesis models: Some solutions. *Quart. J. Roy. Meteor. Soc.*, **97**, 139–153.

Palmén, E., and C. W. Newton, 1969: *Atmospheric Circulation Systems*. Academic Press, 603 pp.

Reiter, E. and J. D. Mahlman, 1965: Heavy radioactive fallout in the southern United States, November 1962. *J. Geophys. Res.*, **70**, 4501–4520.

——, D. W. Beran, J. D. Mahlman and G. Wooldridge, 1965: Effect of large mountain ranges on atmospheric flow patterns as seen from TIROS satellites. Rep. No. 2, Project WISP, Atmos. Sci. Pap. No. 69, Colorado State University, 111 pp. [NTIS PB 168-604].

Shapiro, M. A., 1978: Further evidence of the mesoscale and turbulent structure of upper level jet stream-frontal zone systems. *Mon. Wea. Rev.*, **106**, 1100–1111.

Simmons, A. J., and B. J. Hoskins, 1979: The downstream and upstream development of unstable baroclinic waves. *J. Atmos. Sci.*, **36**, 1239–1254.

EXCERPTS FROM: SATELLITE INTERPRETATION
THE THIRD WEATHER WING/TECHNICAL NOTE - 81/001
EUGENE M. WEBER AND STEVEN WILDEROTTER 28 DECEMBER 1981

Ridges:

Upper ridges in mid-latitudes can best be located by using jet stream cirrus shields. The width of the cloud pattern, its orientation and the characteristics of its forward edge can be used to determine the amplitude or sharpness of the ridge. Upper ridges can be categorized into three main groups consisting of sharp, medium and broad ridges.

• Sharp Ridges - A sharp ridge has a narrow cloud band, a north to south orientation and the cloud system has a sharp leading edge (little, if any, spillover of clouds over the ridgeline; Figure 102). Sharp ridges will be relatively narrow due to the close spacing of the trough and ridge which results in a narrow area of upward vertical motion. Consequently, the leading edge of the cloud band will end abruptly at the ridgeline due to the rapid change from upward to downward motion. A sharp ridgeline pattern is shown in Figure 103; the related IR photo, Figure 104, shows the forward edge of the cloud system ending at the ridgeline.

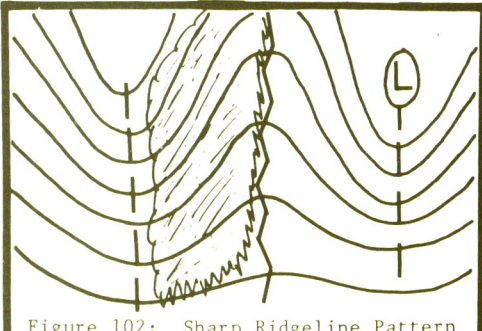

Figure 102: Sharp Ridgeline Pattern

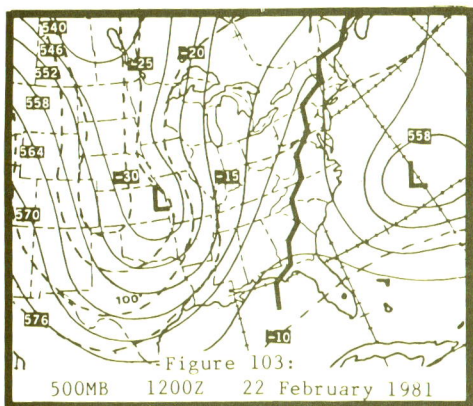

Figure 103: 500MB   1200Z   22 February 1981

Another sharp ridgeline pattern is shown in Figure 105; this pattern is associated with blocking systems. This pattern occurs more often over the eastern Pacific during the winter months when a major ridge builds and persists along or off the west coast of North America and northward into Alaska. Figures 106 and 107 illustrate this event. Trough systems approaching from the west decelerate as they approach the blocking ridgeline. The cloud system becomes stationary and may linger for several days or longer until the blocking ridge flattens or shifts to another area. In these patterns, short waves either move northward west of the ridge towards Alaska or move southeastward towards California (eventually eroding the southern portion of the blocking ridge.) Sharp ridgelines are generally associated with slow-moving meridional trough/ridge systems.

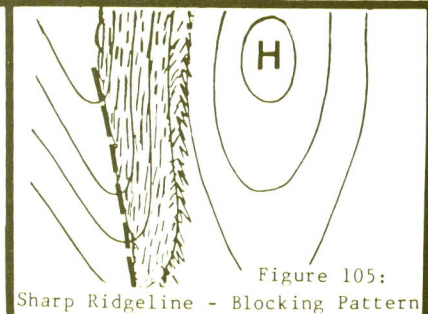

Figure 104: 1130Z   22 February 1981

Figure 105: Sharp Ridgeline - Blocking Pattern

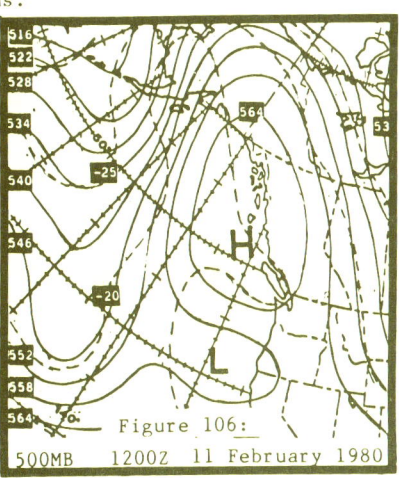

Figure 106: 500MB   1200Z   11 February 1980

Figure 107: 1215Z   11 February 1980

• **Medium Ridges** - A ridge of moderate amplitude has a wider cloud band with a less distinct forward edge and will not end as abruptly as with the sharp ridge (Figure 108). The change in direction of the vertical motion at the ridgeline will be gradual thus allowing clouds to spill over the ridgeline. The cirrus shield extends a few degrees downstream from the ridgeline with the front edge of the middle cloud layer (if it can be seen) corresponding to the ridgeline. Figure 109 depicts a medium ridge pattern; the related satellite photo, Figure 110, shows clouds extending beyond the ridge axis. Of the three ridge groups, medium ridges are observed the most often over the U.S., and are generally associated with transitory short wave systems.

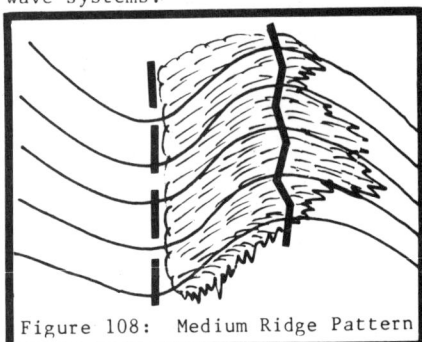

Figure 108: Medium Ridge Pattern

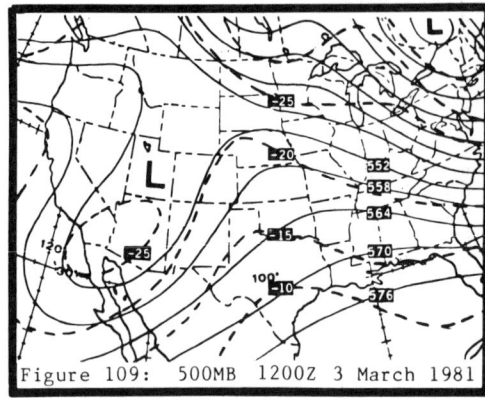

Figure 109: 500MB 1200Z 3 March 1981

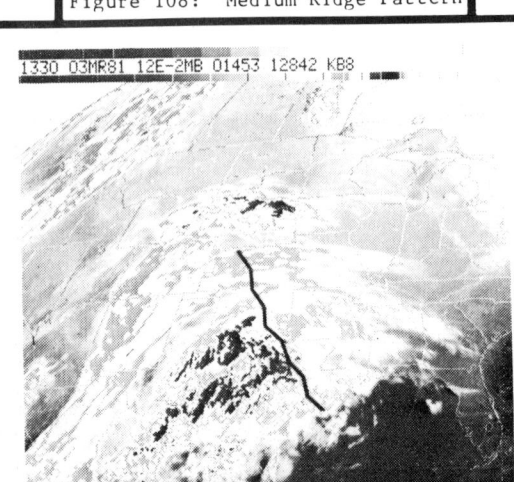

Figure 110: 1330Z 3 March 1981

• **Broad Ridges** - A broad ridge has a very wide cloud band with an almost west to east orientation (Figure 111). A large region of upward vertical motion occurs between the widely spaced trough and ridge. The cirrus shield will extend well downstream from the ridgeline with the middle clouds also extending past the ridgeline. Anticyclonically-curved striations are frequently seen in broad cloud patterns; one of the best methods to locate the exact ridgeline is to use the curvature of the cirrus shield as shown and noted by the arrow in Figure 112. Figure 113 depicts the related 500mb analysis.

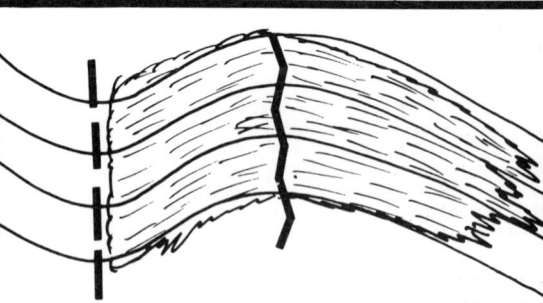

Figure 111: Broad Ridge Pattern

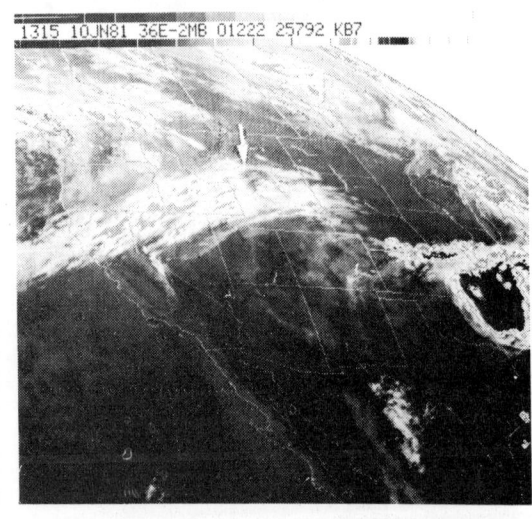

Figure 112: 1315Z 10 June 1981

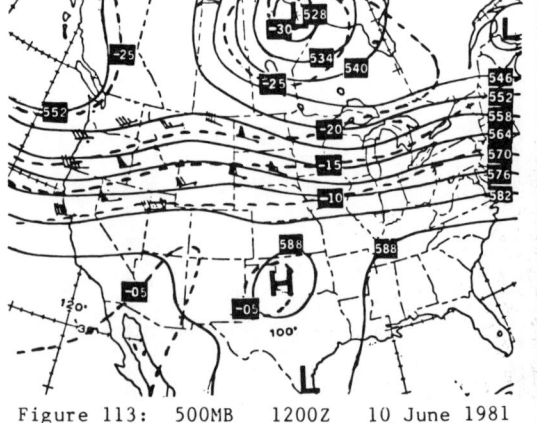

Figure 113: 500MB 1200Z 10 June 1981

2-D-2

Perhaps the most beneficial information that can be obtained from following cloud systems at the ridge line is determining changes in the sharpness of ridges between periods of upper air analyses. In Figure 114, a model of a low amplitude ridge is shown with clouds spilling over the ridge line. Figure 115 illustrates a model for high amplitude ridge (sharp, strong ridge line). Forecasters can follow continuity of the ridge sharpness to determine if a low is intensifying or if it is beginning to break down a blocking ridge pattern. If continuity showed evolution of a cloud sequence such as that in Figure 114, then later like Figure 115, the trough is deepening and the ridge is building. The opposite would be true if the trend was reversed.

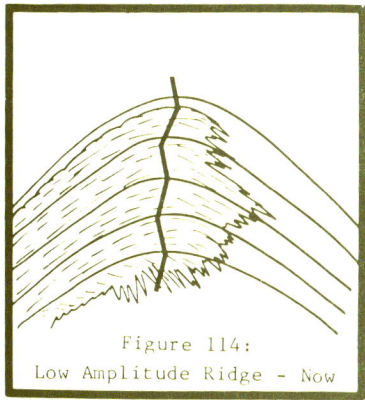

Figure 114: Low Amplitude Ridge - Now

Figure 115: High Amplitude Ridge - Later

Troughs:

The change in the sign of the vertical motion and wind directions at the troughline is reflected on satellite pictures by differing cloud patterns. Three of the most common cloud patterns used to identify the location of mid and upper troughs are: breaks in frontal bands, enhanced cumulus clouds and comma-shaped clouds. These three patterns and combinations thereof should aid forecasters in locating most trough systems.

• Frontal Cloud Bands - The mid tropospheric troughline can frequently be located where a frontal band and trough intersect as shown at point D in Figure 116. At this point of intersection, the frontal clouds will lessen, become fragmented or disappear where the mid level vertical motion changes from upward to downward at the troughline. Figures 117 and 118 illustrate cloud pattern changes (points D) where troughlines intersect frontal bands. East of the troughlines and along the active bands, cloud tops reach into the mid and upper levels reflecting upper vertical motion (and precipitation). West of these troughlines, frontal bands become inactive, change cloud character and decrease in coverage; the clouds are primarily low clouds reflecting mid and upper level downward vertical motion (little, if any, precipitation).

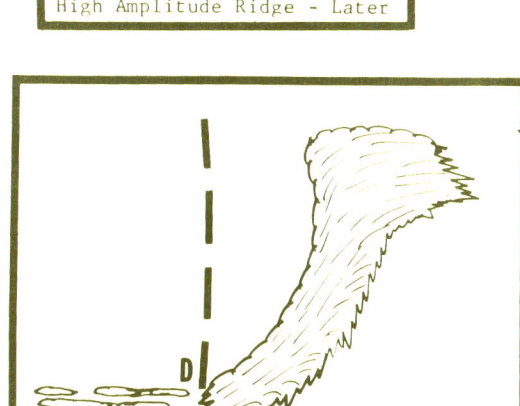

Figure 116: Troughline Identification

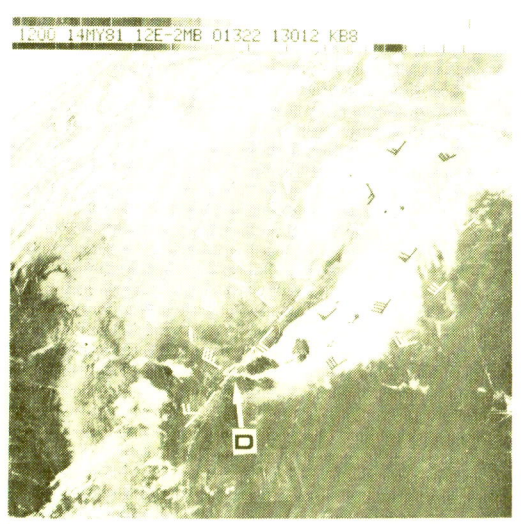
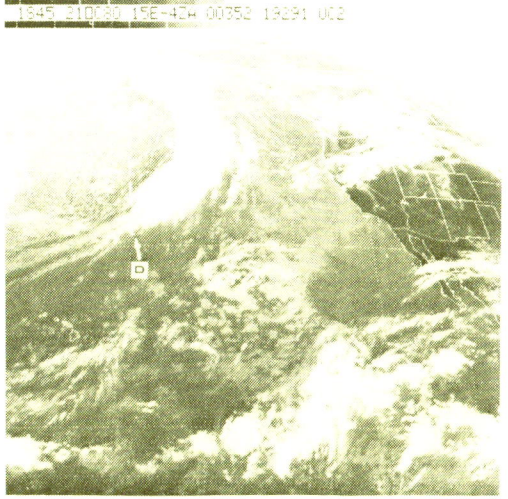

• **Enhanced Cumulus** - Areas of positive vorticity advection are reflected in the appearance of cellular clouds to the rear of cold fronts. The upward motion produces areas of enhanced cumulus. These areas are observed frequently over oceanic areas (and less over land) where there is an abundance of lower level moisture. In Figure 119, an area of enhanced cumulus is noted by the arrow along and to the east of the troughline. The 500mb analysis, Figure 120, (five hours earlier from Figure 119) shows the troughline west of California.

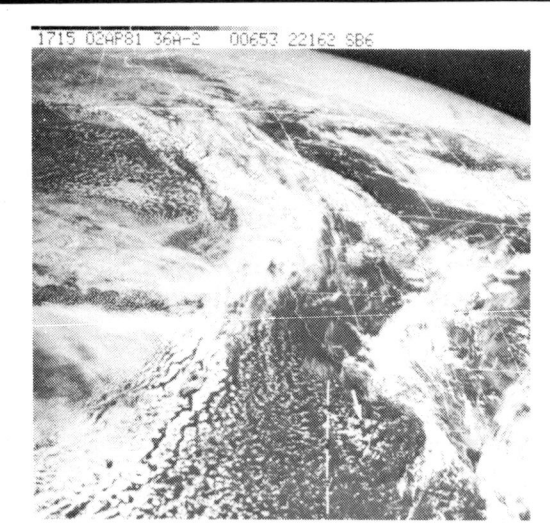

Figure 119:   1715Z   2 April 1981

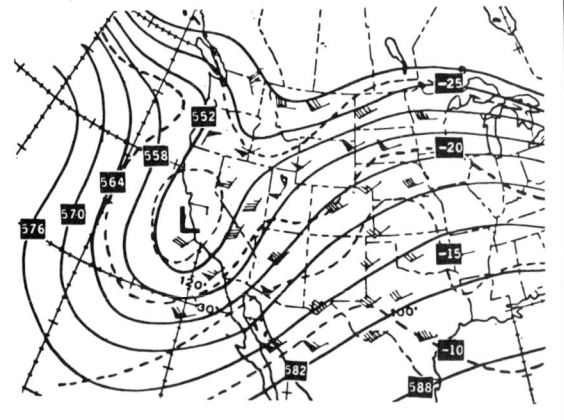

Figure 120: 500MB   1200Z   2 April 1981

Continued vertical development of enhanced cumulus areas (associated with PVA) eventually will form small comma-shaped cloud systems. The anvil plumes reach into the mid and upper levels and form comma-shape cloud systems. This is more likely to occur over land areas during the warm season; Figures 121 through 123 depict such an event over land. In Figure 121, an area of enhanced cumulus appears over most of Mississippi and is noted as area G. The troughline associated with this short wave system is also shown and is placed to the left of the enhanced cumulus area. Further to the east over Alabama, a comma-shaped cloud system (area H) has developed from enhanced cumulus. The related IR photo, Figure 122, shows the vorticity comma cloud system very well (see arrow). The related 500mb analysis (Figure 123; seven hours earlier) shows the short wave trough system.

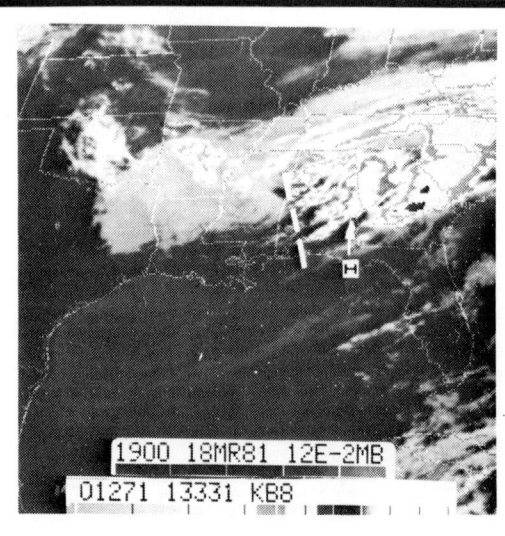

Figure 121:   1930Z   18 March 1981

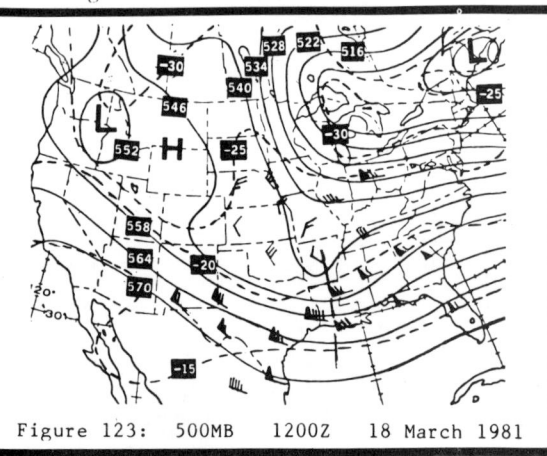

Figure 123: 500MB   1200Z   18 March 1981

Figure 122:   1900Z   18 March 1981

2-D-4

Enhanced cumulus areas located well to the rear of a short wave comma cloud system often reflects either the major trough's position or the presence of a new vorticity center within the cold air of the trough system. An example is shown in Figures 124 and 125. In Figure 124, a short wave comma cloud system is shown over the eastern U.S. The thick cloud frontal band stretches southwestward across southern Georgia and into the Gulf of Mexico. The band becomes fragmented at point F reflecting a change from upward to downward vertical motion in the middle to high levels. The short wave's troughline can be placed at point F; however, when looking at the 500mb analysis shown in Figure 125, forecasters would be hard pressed to locate the troughline. The main troughline and cold air is located further to the west across the central and southern plains as revealed in Figure 125. In Figure 124, the enhanced cumulus area noted at location G over the central Great Plains has developed within the cold air pocket of the major troughline. Forecasters should be alerted, especially during spring (winter), for possible cold air thunderstorms (snow showers) developing in these enhanced cumulus areas.

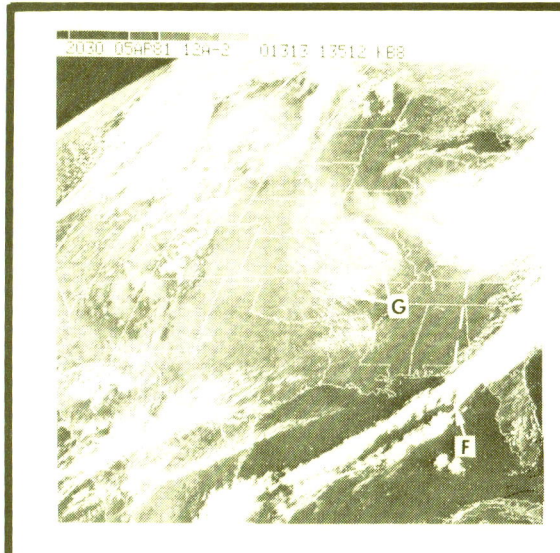

Figure 124:   2030Z   5 April 1981

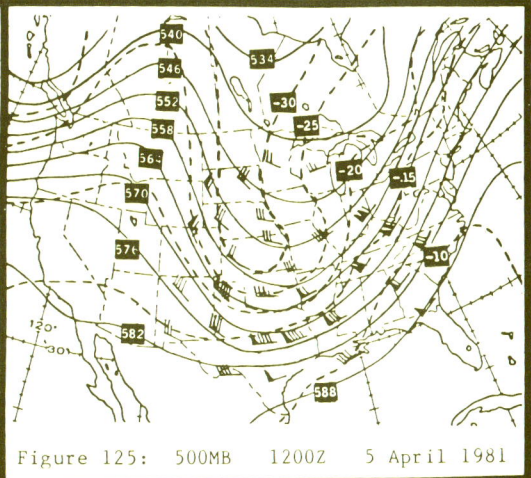

Figure 125:   500MB   1200Z   5 April 1981

• Other Comma-Shaped Cloud Systems - In the preceding discussion it was shown that vorticity comma cloud systems often develop within enhanced cumulus areas. Mid tropospheric level troughlines can also be located using other vorticity comma shape cloud patterns. Over land areas during the cold season, many vorticity comma cloud patterns to the rear of major cyclones are mid-level features and appear to be stratiform in satellite pictures. These systems, however, are often highly convective, but they generally do not produce anvil plumes. Figures 126 and 127 depict two such events.

Figure 126:   1200Z   29 March 1981

In Figure 126 (500mb; 1200Z wind data has been added), a vorticity comma cloud system is noted at location B, and it is located to the rear of the larger comma system. The troughline can be placed to the rear of the comma cloud at location B and also to the rear of the small mid-level cloud system over southern New Mexico. The IR temperature scale indicates that the vorticity comma is primarily composed of mid-level clouds.

A second example is shown in Figure 127. The vorticity comma head is noted at B; the comma tail extends southeastward across southeastern Missouri. A negative-tilt (oriented northwest-southeast) short wave trough can be placed to the rear of the comma cloud. The related 500mb analysis, Figure 128, shows the minor short wave within the flow of the long wave trough system. Negative-tilted short wave systems such as shown in Figure 127 are often dynamic systems and produce strong cyclones.

The tail end of the short wave troughline in Figure 127 is placed at the bend in the frontal cloud band noted at point E. This bend reflects the area of maximum cyclonic curvature upwind of the inflection point shown at point F. Returning to Figure 128, a jet maximum axis of 90 knots across northeastern Texas into central Arkansas can be seen; this jet maximum is punching northeastward and is reflected by the bend in the cloud frontal band.

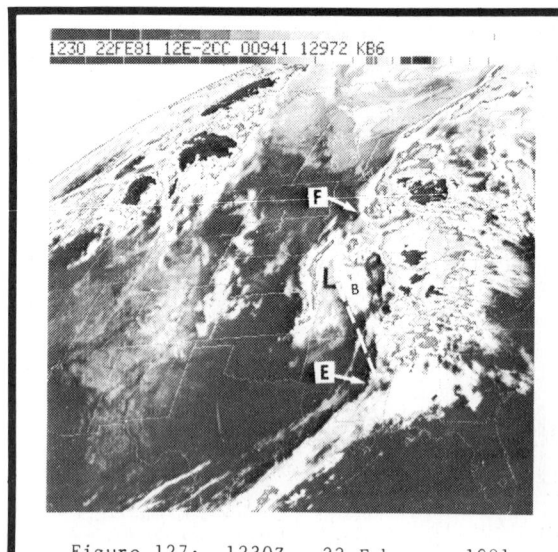

Figure 127:   1230Z   22 February 1981

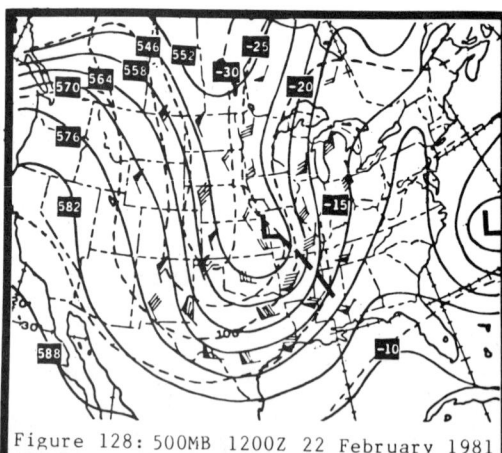

Figure 128: 500MB 1200Z 22 February 1981

- Cirrus Streaks - Forecasters should evaluate all cloud systems for hints in determining troughline positioning. Cirrus streaks may be helpful; this is shown in Figure 129. In Figure 129, a west to east cirrus streak is noted across the Gulf of Mexico. A bend in the streak is shown at point T; this bend reflects maximum cyclonic curvature. Three features in Figure 129 (two have been presented earlier) can identify the general location of the troughline. First, the troughline lies west of the vorticity comma cloud noted at S (hard to see). Secondly, the troughline can be placed in the cyclonic curvature of the cirrus streak. Finally, the cloud frontal band terminates at point U reflecting a change in mid and upper level vertical motion.

Northeast-Southwest Aligned Troughs:

The situations described earlier in troughline identification primarily exist with troughs oriented north to south. There are those occasions, however, when troughs become oriented northeast to southwest (positive tilt) as do the cloud patterns accompanying them. In these cases, the upper level trough does not actually intersect the frontal cloud band so the strong change in cloud character is not evident. Figure 130 is an example of such a system. In these cases, clouds to the northwest are suppressed due to downward vertical motion, but convection, overrunning and low-level convergence team up to produce clouds on the right side of the troughline. In Figure 130, the chance for frontal wave development increases in area D where the cloud frontal band lies parallel to the trough axis. Examples of this type of trough orientation are illustrated in Figures 131 through 134.

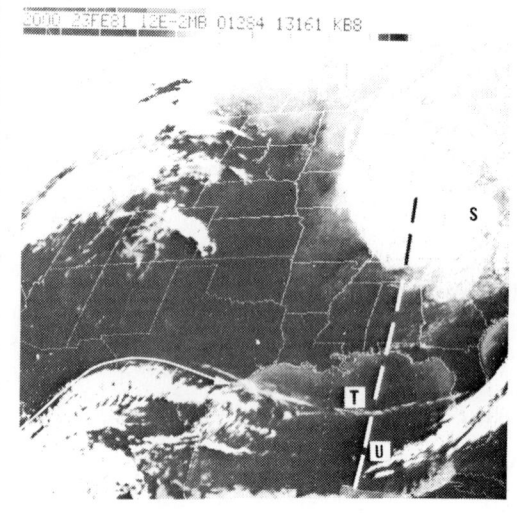

Figure 129:   2000Z   23 February 1981

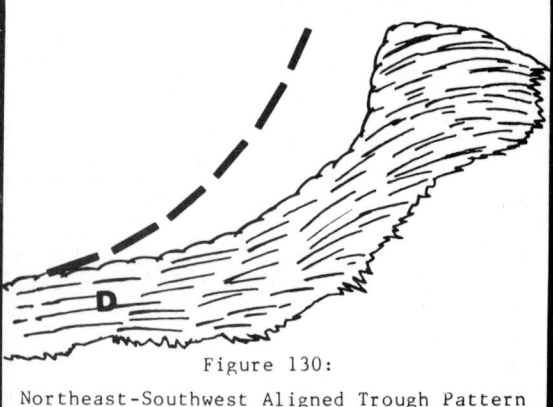

Figure 130: Northeast-Southwest Aligned Trough Pattern

• Example 1 - In Figure 131, the 500mb analysis shows a short wave system moving across the central and upper Midwest. A closed low appears over Mexico. The short wave trough's southern location across Texas is becoming oriented northeast to southwest. In the related visible satellite photo, Figure 132, a cirroform layer is evident across the southern plains (point G) and is spreading northeastward up the frontal cloud band. The cirrus layer is a result of convection which developed across central Texas ahead of the closed low.

• Example 2 - During the warm season, thunderstorm activity often breaks out (especially at night) along and to the right of upper troughlines oriented northeast to southwest. The 500mb analysis, Figure 133, reveals a trough system oriented northeast to southwest across the southern plains. In Figure 134, widespread cloudiness with some thunderstorm activity is clearly evident across Texas and Mexico.

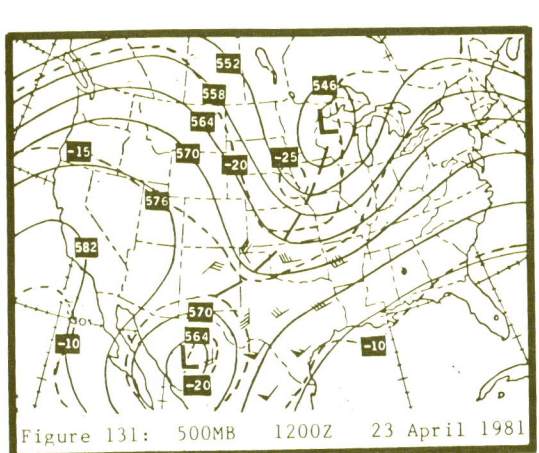

Figure 131: 500MB 1200Z 23 April 1981

Figure 132: 1400Z 23 April 1981

Figure 133: 500MB 1200Z 5 July 1981

Figure 134: 1200Z 5 July 1981

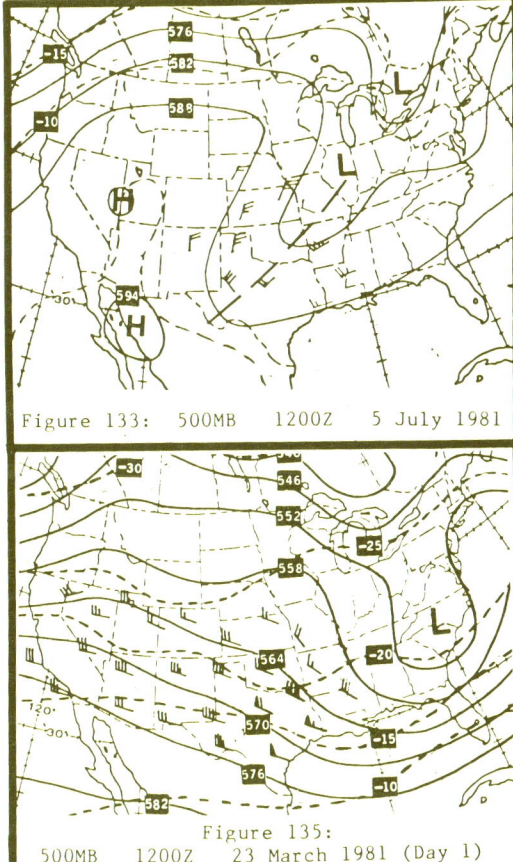

Figure 135: 500MB 1200Z 23 March 1981 (Day 1)

Trough Deepening:

The following example depicts trough deepening within an east to west upper level west of the Rocky Mountains. On Day 1, Figure 135, a west to east 500mb flow prevails over the western U.S. No significant trough system is approaching the West Coast. Some slight thermal troughing is noted over the Rockies.

On Day 2, Figure 136, a trough is evident across Utah and Arizona. The related satellite photo, Figure 137, identifies the trough system very well. During this early stage of development, the troughline can be placed immediately behind the cloud system as indicated in Figure 137. The surface pattern is shown in Figure 138; no surface frontal system appears in the area of the deepening upper trough.

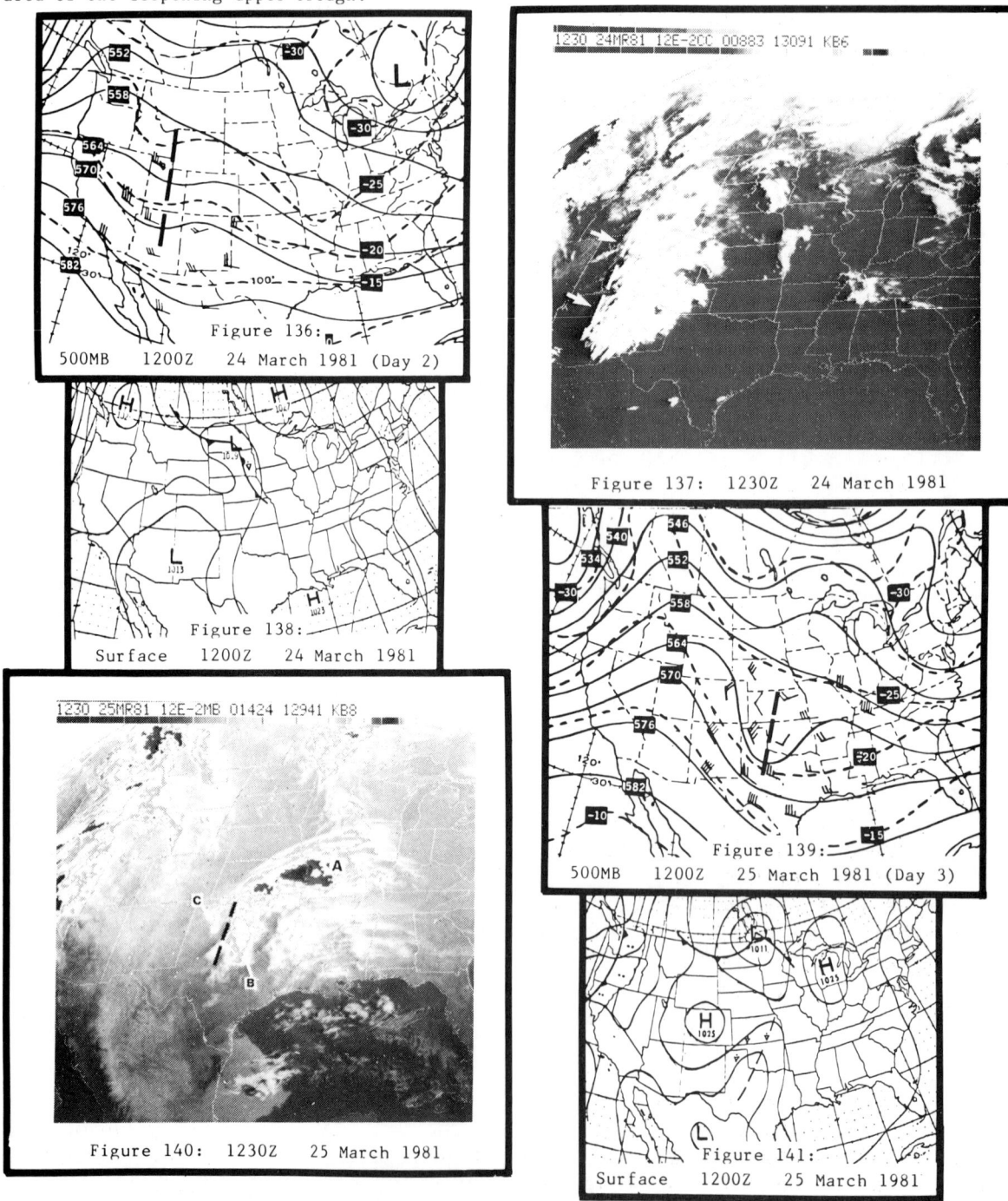

Figure 136: 500MB 1200Z 24 March 1981 (Day 2)

Figure 137: 1230Z 24 March 1981

Figure 138: Surface 1200Z 24 March 1981

Figure 139: 500MB 1200Z 25 March 1981 (Day 3)

Figure 140: 1230Z 25 March 1981

Figure 141: Surface 1200Z 25 March 1981

By Day 3, Figure 139, the trough has continued to deepen and has moved out of the Rocky Mountains. A jet maximum (60 knots) appears at the base of the trough. In the satellite photo, Figure 140, some cyclonic organization is beginning to take shape. The vorticity cloud system is noted at location B, but it has not yet assumed a comma-shape appearance. The troughline lies just west of the vorticity cloud system as shown in Figure 140. A narrow band of deformation zone clouds is noted at location C which may give a hint that a low could subsequently develop within the base of the trough over Oklahoma.

The surface analysis, Figure 141, depicts an inverted trough pattern across the central and southern plains - a reflection of the deepening upper trough. This example typifies those situations when cold fronts suddenly develop east of the Rocky Mountains. In addition, explosive cyclogenesis can occur in the northern portion of the inverted trough below the upper level trough system.

SURFACE CYCLOGENESIS AS INDICATED BY SATELLITE IMAGERY

Frank J. Smigielski

Gary P. Ellrod

Satellite Applications Laboratory (NESDIS)
National Oceanic and Atmospheric Administration
Washington, D.C.

ABSTRACT

Indications of extratropical cyclogenesis which can be observed in infrared, visible or 6.7μm moisture channel satellite imagery are summarized. These include semi-quantitative techniques for estimating the minimum central pressure and the location and strength of maximum surface winds accompanying the low. Satellite analysis of extratropical storm systems is invaluable over oceanic regions where conventional data is limited.

1. Introduction

There are two major challenges facing meteorologists engaged in the analysis and forecasting of extratropical low pressure systems over oceanic areas. The first is determining the location and central pressure of the system at the time of analysis and the second is estimating the movement and amount of development that is expected in a certain time interval. These two factors are especially important in the initial stages of a rapidly intensifying surface low. There are other considerations, such as wind speeds and wave heights which can be ascertained to some degree from knowledge of the depth of the surface low pressures.

The problems confronting the marine forecaster are: first, the lack of available surface ship reports at certain times of day, especially during the nighttime hours; and second, the analyst has to make a decision on whether certain ship reports are valid. The latter is certainly a key factor when observations from a particular ship are missing for several observational periods, then received with a large pressure difference from the one last reported. The tendency is not to believe the report; hence, some good ship reports have been discarded when in reality the data could have been used to good advantage. Even with a normal amount of surface ship observations, cyclogenesis does occur in areas not covered by the reports. Usually there is some indication that cyclogenesis is occurring; however the exact location of the development and the central pressure have to be estimated.

2. Satellite Indicators of Cyclogenesis

Satellite data can be a valuable tool to meteorologists in detecting extratropical cyclones, monitoring their development and estimating their central pressures. This is not a new concept. Early research on the use of satellite imagery as an aid to surface analysis and forecasting expounded the use of satellite observed cloud patterns in the interpretation of surface meteorological conditions, especially the positioning of surface fronts and lows, and the likelihood of cyclogenesis. McClain and Broderick (1967) developed a group of schematics for different stages of cyclone development. The stages pertained to development of low level baroclinic cyclones whose chief characteristic was that development first occurred in the lower levels, then extended upward through the middle and upper troposphere.

Anderson et al. (1973) went one step further in detailing schematics of baroclinic developments by showing the characteristic cloud patterns at about 24 hour intervals. Thus it was now possible to extrapolate a cloud pattern from today's image to show the pattern as it would appear on the following day (Figure 1). In their report, cyclone development was based on: (1) the increase in the width of a cloud band associated with the warm front; (2) an increase in the vertical thickness of the clouds in this same region as revealed in either the infrared or visible data; (3) the appearance of anticyclonic curvature of the main cloud band on the cold air side; or (4) the appearance of mesoscale streaks of upper and mid level cloud along the anticyclonically curved boundary of the cloud mass, indicating that the process of mass outflow has begun in the upper troposphere. These are indications that a closed surface low should be expected in the next 24 hours. It was now possible to provide meteorologists with cloud patterns indicating cyclogenesis and subsequent cloud patterns at 24 hour intervals. These patterns were later modified for Southern Hemisphere use by Guymer (1978).

In addition, Anderson et al. introduced another form of cloud pattern, which was referred to as a barotropic positive vorticity advection (PVA) cloud pattern. It was shown that when this pattern interacted with a baroclinic cloud band, cyclogenesis often occurs. When a comma cloud, reflecting a maximum of positive vorticity advection approaches a frontal band, wave formation begins. When the comma cloud continues to intensify, cyclonic circulation increases at low levels, resulting in a more northerly flow in the cold air behind the vorticity center and more southerly flow of relatively warm air ahead of the PVA. This effectively creates a surface cold front between the cold and warm air. When the comma cloud merges with the baroclinic band it gives the appearance of an occlusion on the frontal band. The completed system is actually the merging of two separate systems.

Smigielski (1973) introduced a series of developmental and non-developmental cloud patterns for the daily use of the Synoptic Analysis Branch of the National Environmental Satellite Data and Information Service (NESDIS) in briefing the National Meteorological Center (NMC) on Northern Hemisphere surface features. This classification of systems was obtained through daily inspection of cloud photographs on a global basis. It was here the discovery

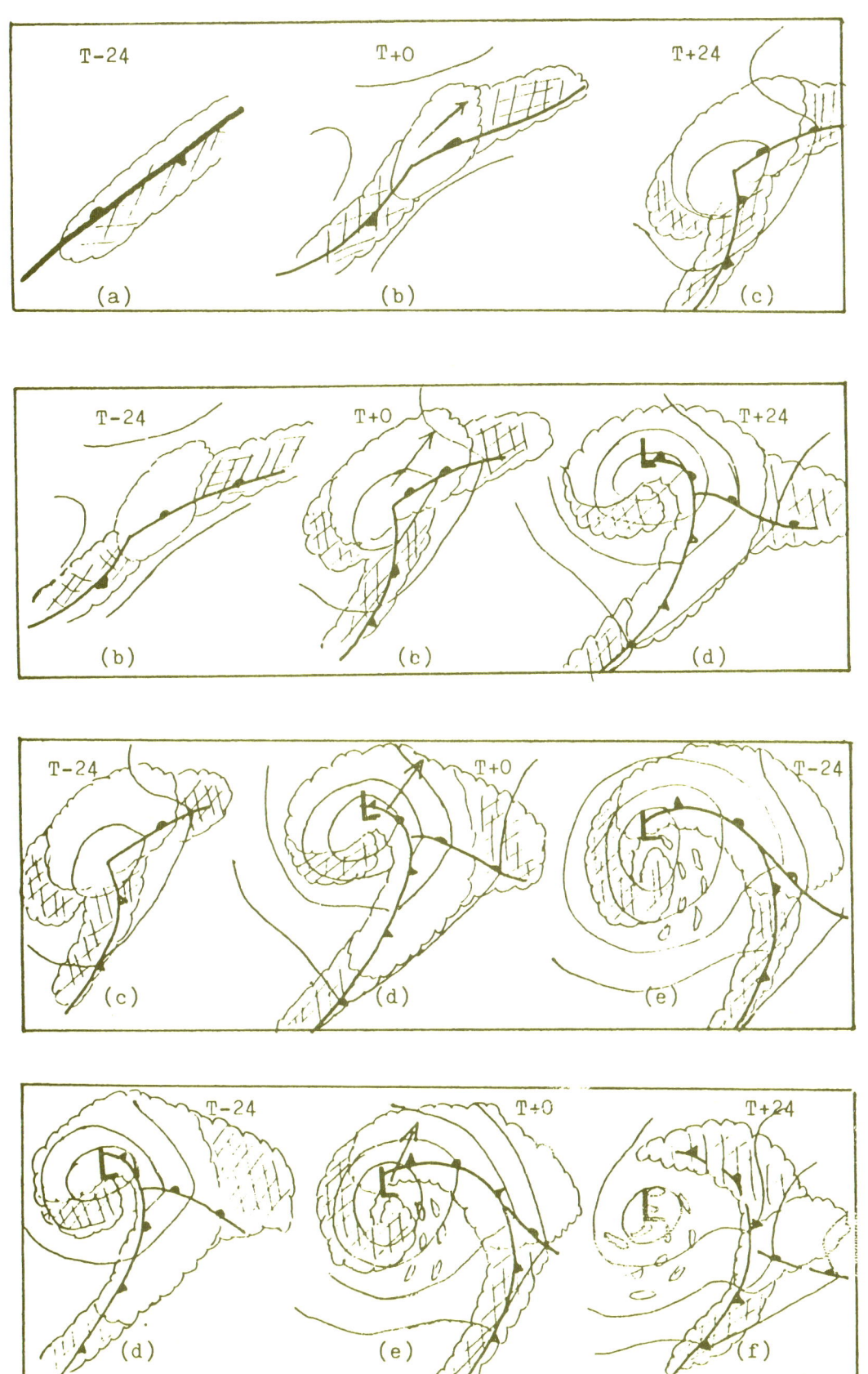

Figure 1. Cloud pattern evolution at 24 hour intervals for an extratropical cyclone developing at a typical rate. (From Anderson, et al, 1973)

was made that most normally developing cyclones go from incipient wave to full maturity in 48 to 72 hour cycles. Rapidly developing cyclones go through this process in 18 to 24 hours (Figure 2). After this period of development a slow dissipation of the cloud system usually occurs. If some reinforcing influences, such as positive vorticity, are injected into the dissipating cyclone, the tendency is then for the cyclone to maintain its lowest pressure instead of filling.

3. Rate of Intensification

In addition to the aforementioned baroclinic and barotropic developments, there are indications whether intensification will occur at normal progression, or whether the development will be rapid. Explosive or rapid deepening occurs whenever a baroclinic cloud band shows a bulging (convex) appearance that is increasing rapidly with time. This is in conjunction with a rapid dissipation of clouds in the cold air behind the baroclinic band resulting in a concave back edge. The combined concave-convex feature results in a characteristic "S" or "C" shape (Younker,1981). A "V" shaped wedge eventually develops along the concave portion of the baroclinic cloud band. This is in response to substantial cold air advection and sinking motions at low and mid levels. When a system begins to develop rapidly, subsidence is large and cumulus development is suppressed within this slot (Burtt and Junker, 1976). Secondly, jet-associated cirrus forms in response to strong PVA to produce a cap on the poleward edge of the main baroclinic cloud system. This cirrus "cap" displays both cyclonic and anticyclonic curvature. (For detailed descriptions of three-dimensional flow in cyclones relative to the comma cloud, see Carlson, 1980 and Weldon, 1979). Pressure rises are characteristic of the dry slot, while falling pressures are found beneath the cloud mass which is expanding with time. Twenty-four hours after these indicators are noted, a tightly wound spiral is visible denoting the cloud circulation center. An excellent example of explosive deepening is shown in Figure 2. The central pressure fell 50 mb in 24 hours to a minimum of 956 mb by 00Z November 14. With systems such as these, minimum central pressures are frequently below 960 mb with sustained winds of 50 to 60kt (25 to 30 m sec^{-1}) observed around the center. At this time, the strongest circulation is still in the lowest levels; however, deepening has occurred to its maximum potential (Jager, 1984). By day 2, or 48 hours from the incipient stage, the system is fully mature with a closed low at 500 mb.

Sanders and Gyakum (1980) have examined the characteristics of a large number of extratropical lows which exhibited explosive deepening (at least 1 mb/hr for 24 hours). Included in their sample was the storm which damaged the luxury liner Queen Elizabeth II and sunk another ship on November 10, 1978. Most of these storms were found to develop in areas of strong sea surface temperature gradients, such as those associated with the Gulf Stream current along the east coast of the United States. Development occurred within or poleward of the strongest westerlies about 400nm (250km) downstream from a mobile 500mb trough.

Burtt and Junker (1976), used SMS (Synchronous Meteorological Satellite) data to describe a rapidly deepening surface low over the eastern Pacific where the pressure dropped from 988 mb to 952 mb in an 18 hour period. Prior to rapid

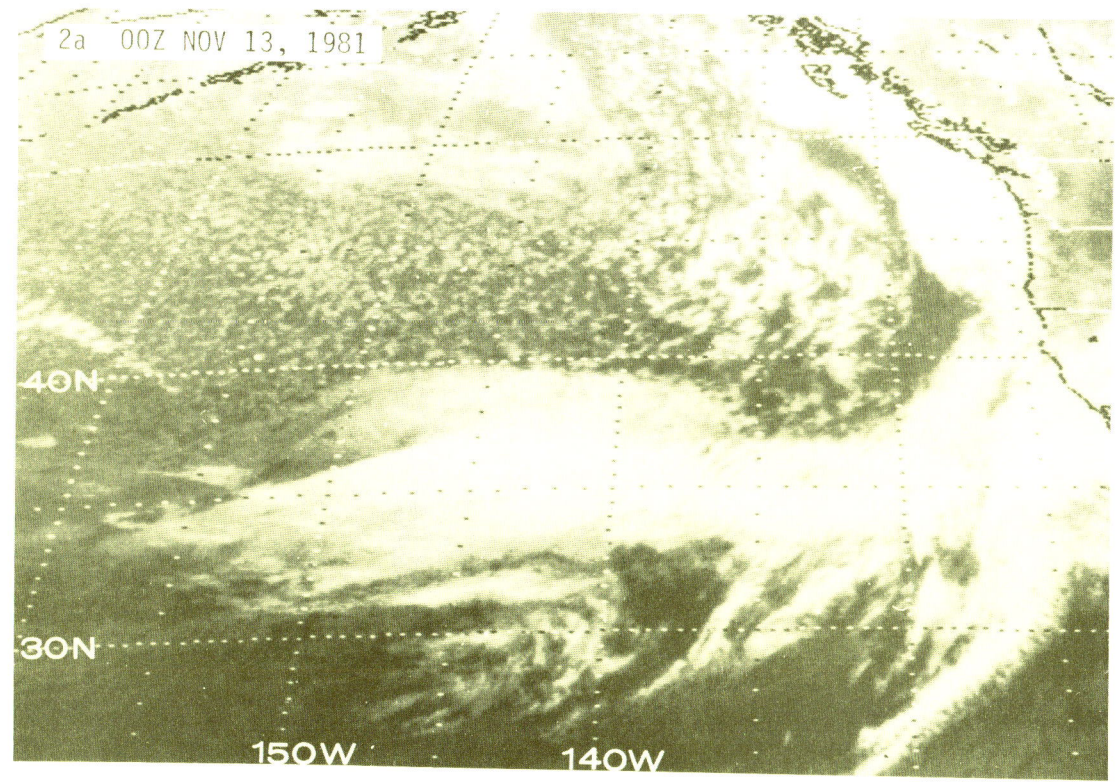

Figures 2a and 2b. Infrared images of a rapidly developing cyclone in the formative stages. (After Jager, 1984)

Figures 2c and 2d. Same storm as 2a and 2b at the mature stage. Storm reached peak intensity in 24 to 36 hours. The central pressure fell 50 mb in 24 hours to a minimum of 956 mb.

deepening, the cloud pattern showed the typical convex, concave pattern and the cirrus shield extended into the cold air from the expanding convex portion of the cloud mass. Younker (1981), using graphs developed by the Anchorage, AK Forecast office used these same guidelines and the initial surface pressure obtained through NMC analyses to arrive at a central pressure to deepening storms reaching the Alaskan coast.

4. Special Types of Cases

A. Triple Point Development

There is a type of surface development that occurs when an occluded system attempts to penetrate a ridge line. In this case (shown schematically in Figure 3), the development occurs at the base of the occluded system near the triple point[1]. The indicators are: (1) a depressing of the baroclinic band just west of the ridge line; and (2) a wind maximum indicated in the satellite imagery by short cirrus bands that approach the base of the occlusion from a westnorthwesterly direction. This wind maximum usually undercuts the back side of the baroclinic band (Weldon, 1979) and indicates the presence of cold air advection needed to begin the cyclogenetic cycle. In addition, (3) the original occlusion and its cloud circulation center will show signs of decay and dissipation. Two of the three indicators should be present for development to occur.

An example of a triple point development is shown in Figure 4. Figure 5 shows the accompanying surface charts for the period 00Z April 3 through 00Z April 6 at 24 hour intervals. The surface data for 00Z April 3, corresponding to Figure 4a has a surface low analyzed near 50°N 160°W. The satellite image for 2015Z April 2 indicates that cyclogenesis should occur near 45°N 160°W. By 2115Z April 3, the satellite image shows a well developed low near 44°N 142°W. On the following two days, this low moved to 31°N 125°W and showed definite signs of development in the cloud pattern. The surface charts for this same period indicate a track to the southeast with some development between 00Z April 5 and 00Z April 6. The deepening cycle was not dramatic because the low was entering an area of normally stable conditions associated with the subtropical high pressure area off the California coast.

B. Low Latitude Cut-off Lows

When a baroclinic cloud band extends south of 35°N latitude and a split develops in the 500 mb flow, a "sub-westerly" cut-off type of development may occur (Figure 6). This type of development usually does not result in a very deep surface system, and at times is reflected by an inverted trough with possibly one closed isobar around the center. These systems are generally very slow moving, they are under the influence of light winds aloft with the main belt of strong westerlies farther to the north.

---

[1] The triple point is the location where the cold front, warm front and occlusion intersect in an extratropical cyclone.

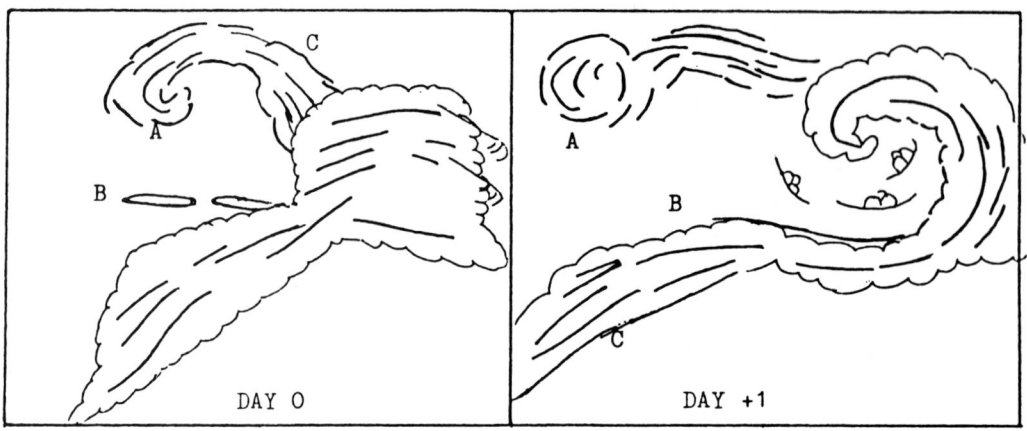

1. The original center shows signs of filling at "A".

2. A polar jet "B" indicated by cirrus streaks cuts across the baroclinic band.

3. The occluded portion of the front "C" shows signs of decay.

1. The old occluded center "A" is filling rapidly.

2. The jet cirrus "B" has a marked cyclonic/anticyclonic appearance on the poleward side of the main frontal band.

3. Usually a stable wave will appear on the frontal band upstream of the main frontal wave ("C").

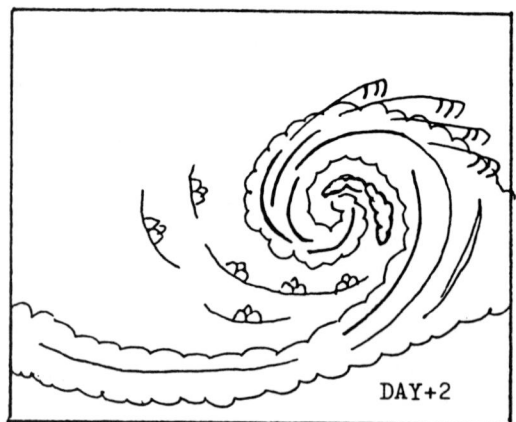

1. Maximum development has taken place.

2. The system can be expected to start a slow filling process.

3. The ridge over which the system was traveling is now fully depressed to the south.

Figure 3. Schematic showing the development of a cyclone at the base of an occlusion (After Smigielski, 1973).

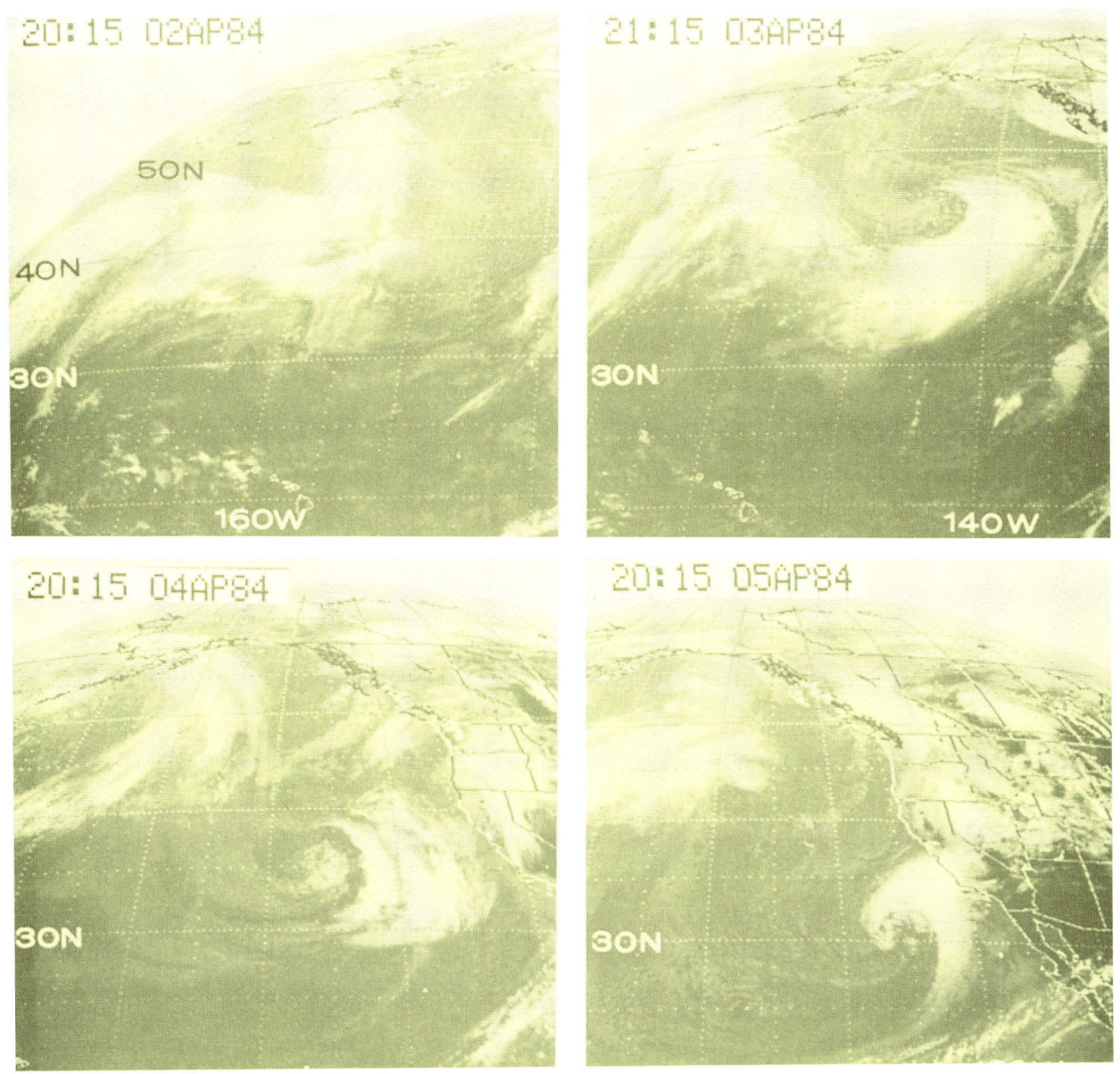

Figure 4. Series of infrared images showing cyclogenesis along an old occlusion.

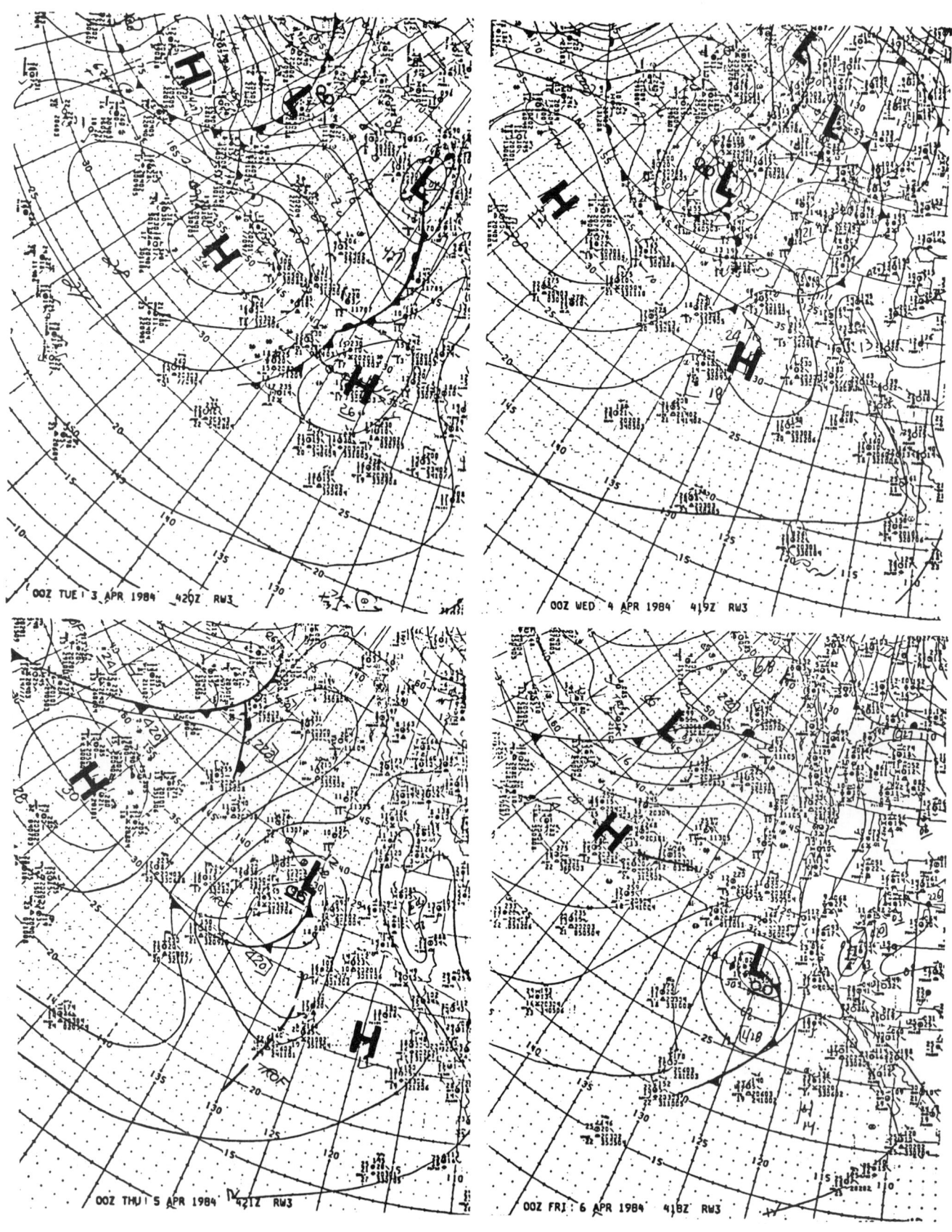

Figure 5. Surface charts at 24 hours intervals for the period 00Z April 3 through 00Z April 6, 1984.

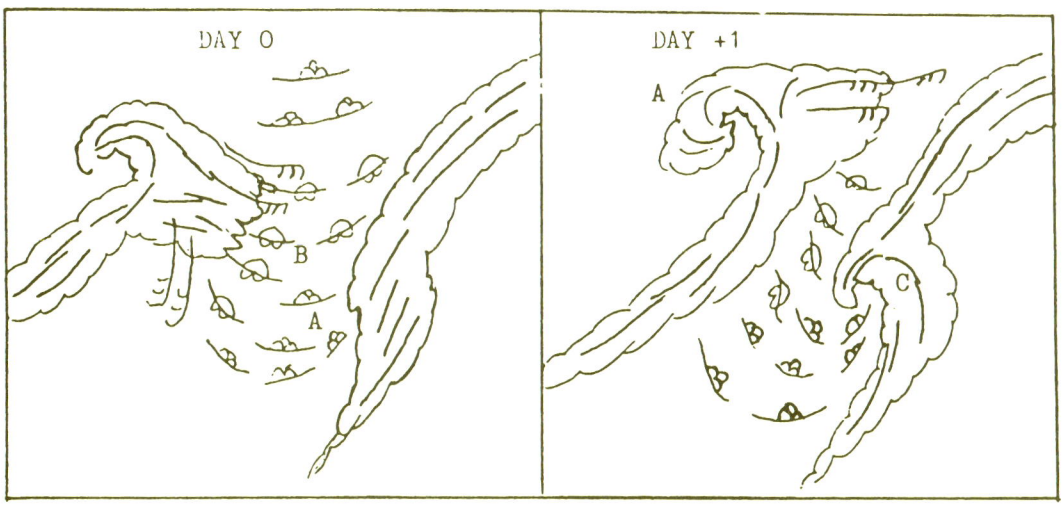

1. The frontal band shows the typical bowing effect at "A".	1. The western system occludes and moves northward at "A".
2. An area of stratocumulus clouds forms between the two systems at "B".	2. The eastern system moves slowly and deepens in the southern portion at "C".
3. A split in the upper trough is indicated by the stratocumulus.	3. The eastern system ("C") is essentially cut-off from the westerly flow.
	4. Maximum development occurs around this time.

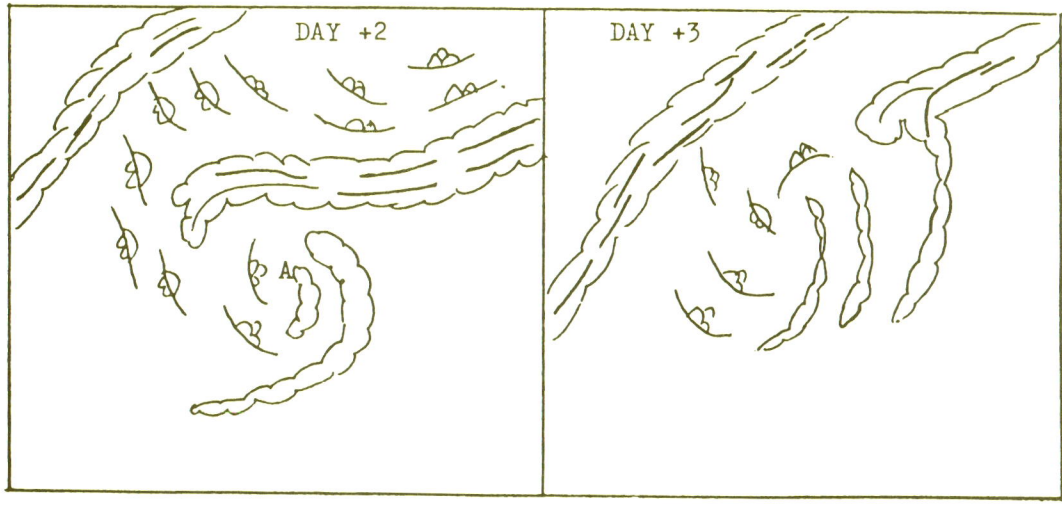

1. The frontal band around the cut-off "A" weakens.	1. The system shows further decay in cloud structure.
2. The primary system is well to the east of the cut-off.	2. Only fragmented cumulus remain around the center.

Figure 6. Schematic of cloud patterns showing a deepening, slow moving cut-off low. (After Smigielski, 1973)

One way a low latitude cut-off low can form is when the northern portion of a trough in the westerlies advances eastward faster than the southern portion. An example of this type of development as it would appear in satellite images is shown in Figures 7a-c. This type of low usually develops first at the upper levels then works down to the surface, rather than vice-versa, as in a normal baroclinic development.

C. Non-Developing Systems

Throughout this paper we have been discussing the indications of cyclogenesis in satellite viewed cloud patterns. There is, however, the case where some indicators exist for surface development but no development occurs. This usually happens when the forward or equatorward side of a baroclinic band has a narrow rope-like line indicating the location of the surface front (Figure 8). This rope-like band is generally more evident in the visible data than in the infrared imagery. Although a concave-convex appearance is indicated in the main cloud field (Figure 9), no development takes place. This type of situation indicates the presence of stable waves along the frontal cloud band with very little if any vorticity advection. The main upper level flow is nearly parallel to the baroclinic band. The band along the cold front portion exhibits breaks or, if overcast conditions occur, the upper levels are composed of thin cirrus. Eventually cyclogenesis may take place; however, the first indication of this would be the elimination of this rope-like line at the leading edge of the baroclinic cloud band.

Generally this non-developing situation exists when a cold front approaches a stable high pressure area located to the south and east of the cold front. The frontal band will exhibit some weak waves; however, the waves will ride around the high without any development on the baroclinic band. Figure 9a through 9c present a stable wave sequence as the baroclinic band near 153°W at 9a moves to near 145°W by 9c. The narrow rope-like along the leading edge of the baroclinic band is visible in all three images. Although waves form along along the band, they all move around the stable high pressure area located in the eastern Pacific. By 0445Z January 28, the baroclinic band has eroded to mostly thin cirrus, with the rope-like line clearly visible from 52°N 125°W to a wave at 45°N 145°W thence to 35°N 146°W.

5. Estimating Central Surface Pressures

Early attempts at estimating the central pressure of extratropical cyclones were made by Troup and Streten (1972). Surface pressure anomalies were determined from cloud patterns in each stage of storm development. Anomaly values were then combined with mean sea level charts to obtain the central pressure of storms.

More recently, Junker and Haller (1980) have developed a system for assigning central surface pressures to low centers based on the amount of circulation visible in the satellite cloud photos (Figure 10). This study was conducted between the months of October and April and was based on developing cyclones in strong baroclinic zones defined by a broad band of multi-layered clouds. This technique should enable meteorologists to better estimate the

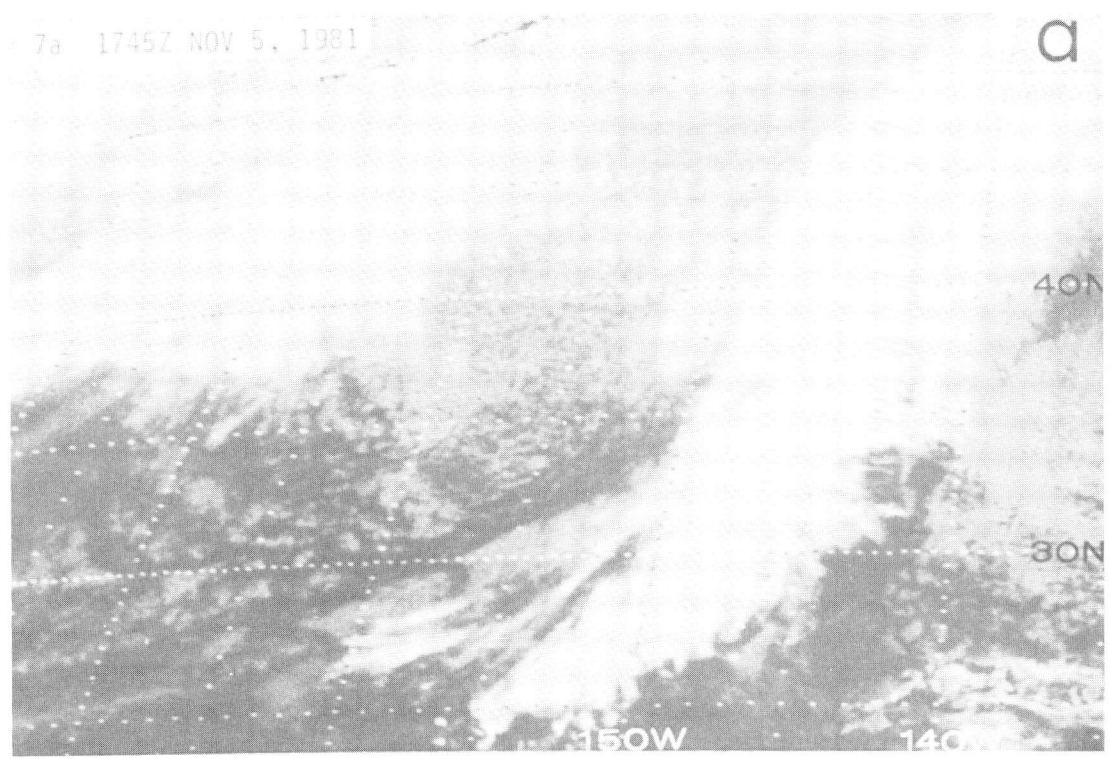

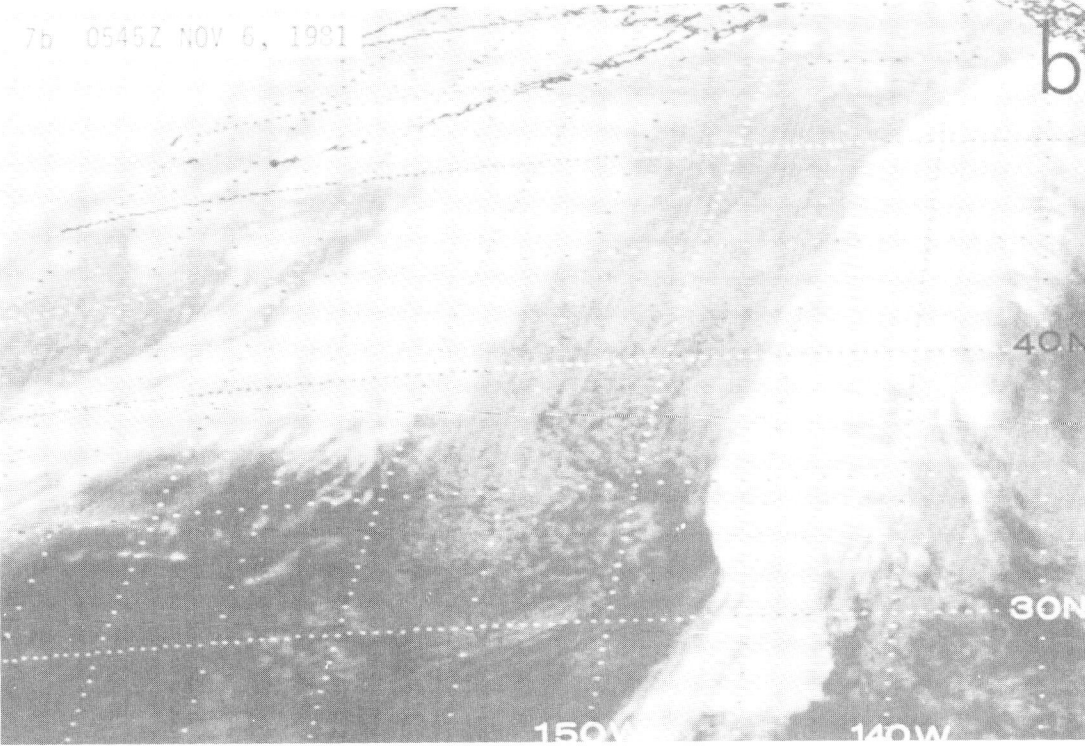

Figures 7a and 7b. Infrared satellite images showing the initial stages in the development of a "sub-westerly" cut-off low pressure system.

Figure 7c. A low latitude development located near 35°N 145°W, 24 hours after Figure 7a.

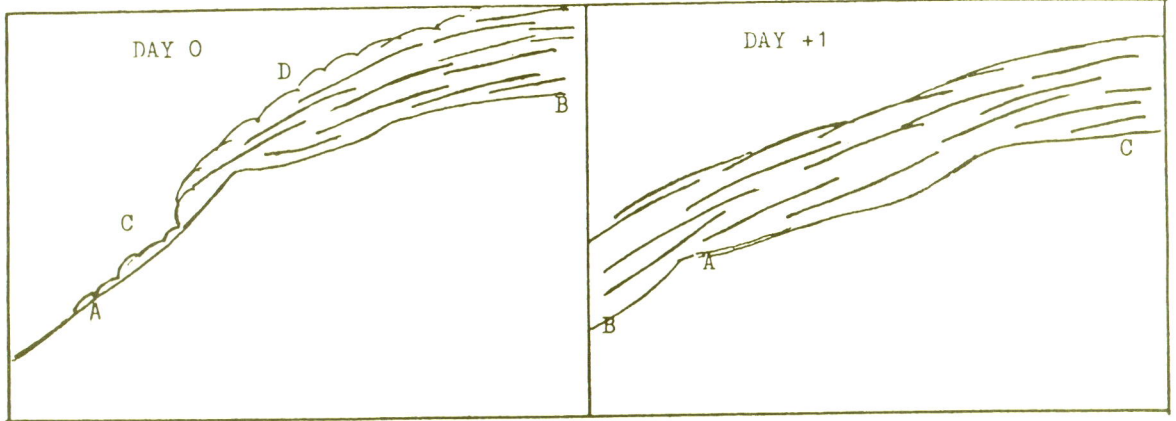

1. The foreward, or equatorward side of the cloud band has a narrow, rope-like appearance at "A-B".

2. The main cloud band exhibits breaks at "C".

3. Cirrus is usually present, however it has a thin appearance.("D")

1. Changes in the appearance of the cloud band are minor.

2. The thin bright line is evident on the equator side from "B" to "C".

3. Another wave may appear upwind from the previous one ("A").

4. The entire band has thin spots.

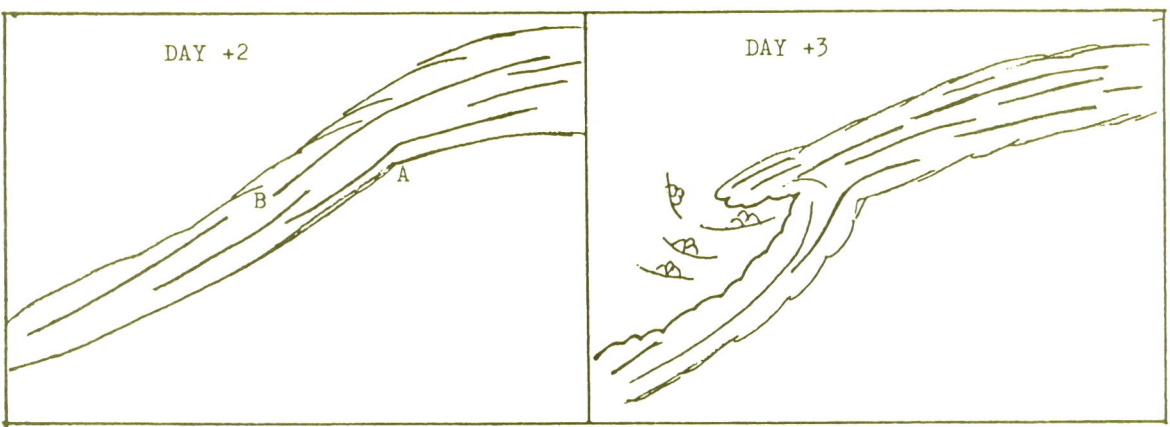

1. A solitary stable wave appears at "A" with cloud striations parallel to the band, within the band.

2. The thin bright line persists.

3. Cirrus is pronounced, but it is parallel to the main cloud mass ("B").

1. If wave development is to begin, the bright line becomes less distinct.

2. The poleward side of the cloud band becomes sharply defined and cumulus begins to intersect the main cloud mass at a sharp angle.

Figure 8. Schematic showing a slowly developing or non-developing baroclinic cloud band (After Smigielski, 1973).

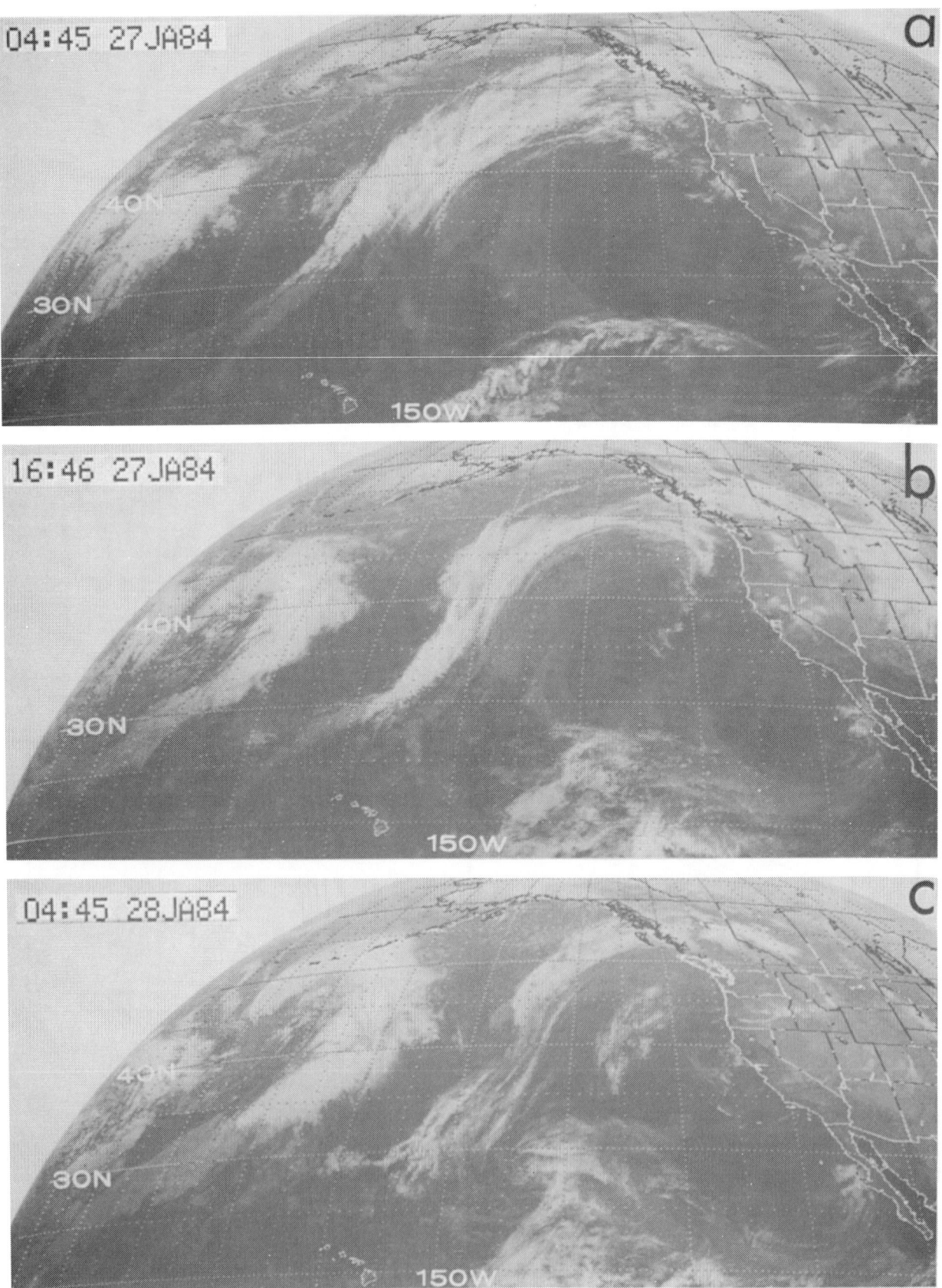

Figure 9a-c. Stable type waves along a baroclinic cloud band (After Smigielski, 1973).

Patterns characteristic of pressures between 990mb-999mb. Note dry slot beginning to form on rear edge of cloud band.

When distinct hooked shaped pattern starts to emerge, pressures dip into 980's.

Figure 10. Estimation of central pressure based on satellite cloud patterns for the developing stage of a cyclone (From Junker and Haller, 1980).

Pressures in 970's indicated by cloud band wrapped around center almost one time.

Pressure near 960mb, cloud band wraps completely around center 1 1/2 times.

Figure 10 Continued. Estimation of central pressure based on satellite cloud patterns for the mature stage of a cyclone (From Junker and Haller, 1980).

surface low pressure within a cyclone in the absence of surface reports. In turn, better wind speed estimates are likely, which should lead to better forecasts of ocean wave heights.

One change to the system as prescribed in Figure 10 is a modification for cyclogenesis when surface pressures are already low when the cyclogenesis process begins. The present system assumes that the surrounding pressures during initial stages of development are above 1000 mb. In the case where existing pressures surrounding the developing system are lower, say around 980 mb, the system calls for lowering each category by a corresponding 20 mb. Therefore, the 970 mb category will then indicate 950 mb, etc.

One other restriction restriction applies to the system. It should be confined to areas north of 40°N latitude. However, this is an arbitrary latitude and may extend farther south if the nature of the cloud pattern suggests a strongly baroclinic system.

6. Location and Intensity of Strongest Surface Winds

The final consideration in surface analysis and forecasting is the question of where the strongest surface winds will be found with respect to the low center and how strong they will be. During specific portions of the development cycle, the strongest winds will be found in different quadrants of the storm, as shown in Figure 11. On the average, the maximum winds during initial stages of development occur in the eastern portion along and ahead of the baroclinic band in the warm air. As the cyclone develops, the strongest winds are found in the northern semi-circle; and finally as the system reaches full maturity, the winds become strongest in the southern semi-circle of the system. As the storm slowly fills, the strongest winds encircle the storm center.

In the case of a "sub-westerly" development, the strongest winds occur across the northern semi-circle of the cloud system. This is usually the case when a blocking high is located north of the closed circulation. However, if the cloud system persists over a long period of time, say greater than 72 hours, then the strong winds will become concentrated across the whole system.

The strength of surface winds can only be estimated from satellite images in the absence of obscuring high cloud layers. The presence of open cellular cloud patterns in satellite pictures has been related to the presence of wind speed of 20 kt (10 m sec^{-1}) or higher (Anderson et al., 1974). When these open cell patterns become deformed into lines, speeds can sometimes exceed 50 knots (25 m sec^{-1}). Satellite-derived wind vectors determined from the motion of low clouds using interactive computers provide wind estimates which are most representative of the 3000 foot (1 km) level (Hubert and Whitney, 1971). These cloud motion vectors may approximate the surface wind speeds in regions of instability where momentum at the cloud level is transported to the surface by strong mixing. Due to multiple layered clouds and high stability, the north or northeast quadrants of extratropical lows are usually not suitable for using cloud motion vectors as estimates of surface winds.

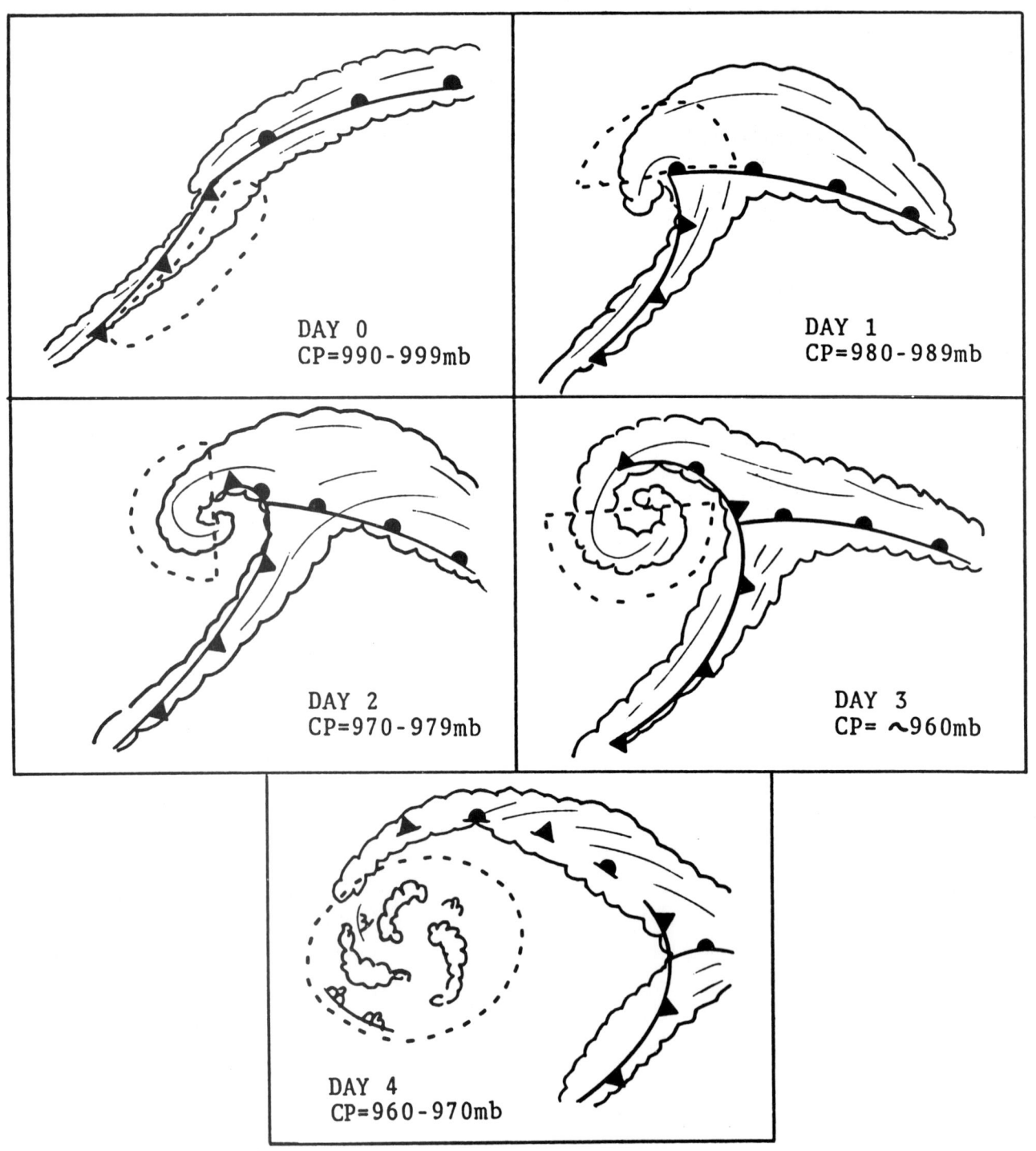

Figure 11. Schematic showing the locations of maximum surface winds (areas outlined by dashed lines) at various stages of cyclone development. Minimum central pressures (CP) are also shown. (From Smigielski, 1973)

In certain extratropical cyclones, characteristics similar to those of tropical storms can be found. In these cases, the strongest winds are located in a relatively small area near the center of the storm (Sanders and Gyakum, 1980). These types of lows often exhibit a spiral cloud structure with an eye-like center (Rasmussen, 1981). Such systems can show explosive deepening, resulting in hurricane-like winds and central pressures. The development of warm core lows from cold core systems often occurs over relatively warm waters such as off the east coast of the United States. The addition of latent and sensible heat from the ocean into the atmosphere is believed to be the cause of this transformation.

Figures 12a through 12d show the progresson of a subtropical development from a predominately cold core in Figure 12a to a warm core system by 12c and 12d. The low center deepened and moved slowly northward, retarded by a strong high pressure ridge which extended from New England westward through the Great Lakes to the southern Mississippi Valley. The minimum central pressure reached 980 mb near the time of the last image, and ships reported sustained winds of 50 kt (25 m sec^{-1}) with seas around 30 ft (9 m).

7. **Use of VAS Moisture Channel Imagery**

Although the 6.7 μm water vapor data is a relatively new tool, some indications of surface development and deepening can be inferred from this information. When an elliptical dark band (indicating the presence of dry air at mid levels) appears behind an "S" shaped pattern in the 6.7 μm data, the associated cloud system (as observed in IR, Vis or 6.7 μm channels) will move rapidly and usually develop until the dark band becomes diffuse. Small dark areas indicating sinking behind baroclinic cloud bands or with PVA comma clouds, are initial indicators of deepening or cyclogenesis. The dark areas will become elongated and larger as the cyclogenesis process continues, usually spreading out along the rear portion of the baroclinic band and spiraling into the vortex center. This usually indicates the end of the deepening process. When the dark band becomes diffuse, some filling of the system will occur.

As previously mentioned, the presence of a dark band behind the front indicates rapid movement. When the band fills in and becomes grayish in appearance, this indicates a deceleration of the front and a slow dissipation of the cloud system.

If the dark area associated with a vorticity maximum or the frontal band is not changing with time, i.e., is not enlarging or not filling in, there will be little change in the associated cloud mass. No increase or decrease in pressure associated with the dark band will be noticed. If the dark area is increasing in area with time, the associated vortex is deepening. Figure 13 shows a cyclogenesis event over the eastern U.S. Figure 13a indicates the first clue that cyclogenesis is occurring. This is indicated by the dark slot located south of Mobile, Alabama, as strong subsidence takes place behind the developing wave. Twelve hours later, the dark band has expanded and a fully developed wave has moved to the Mid Atlantic area. Finally, 13c and 13e show a strong vortex in

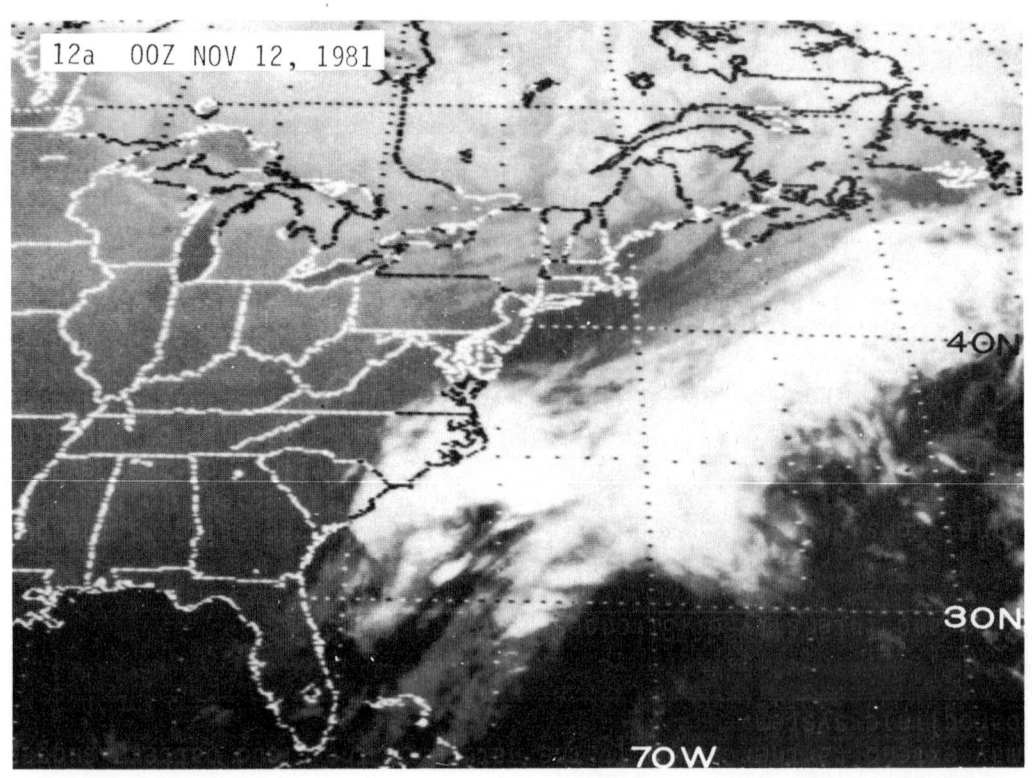

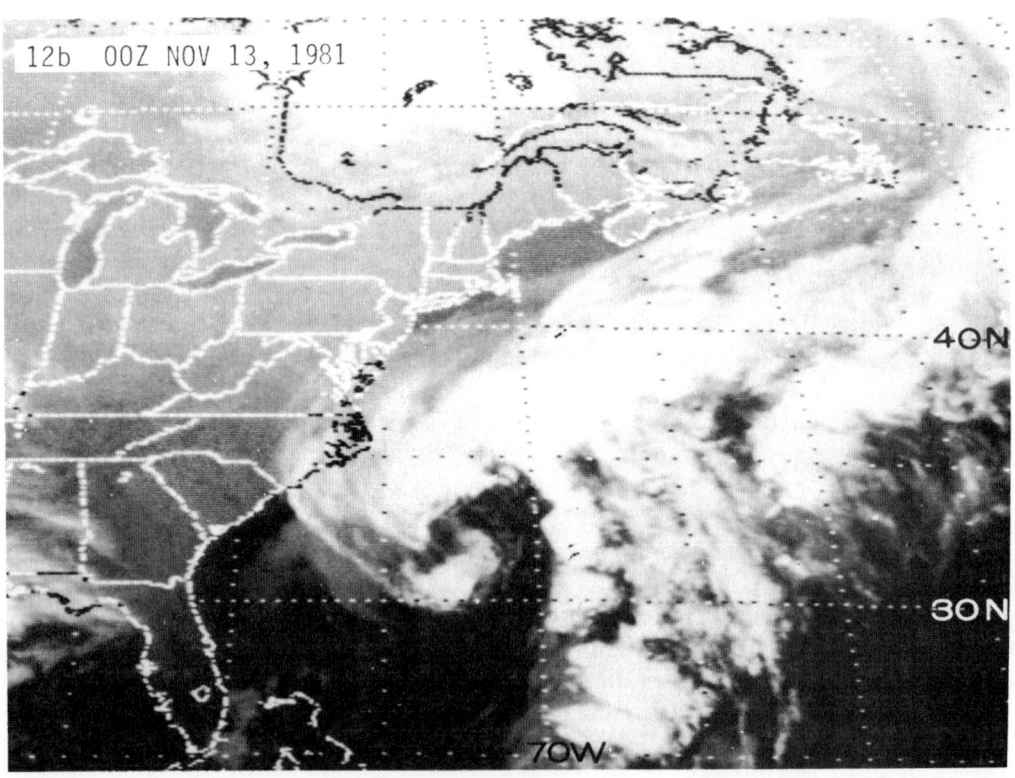

Figures 12a and 12b. A low latitude system acquiring tropical characteristics.

Figures 12c and 12d. The same low latitude system as in Figure 12a and 12b has now reached maximum intensity. The minimum sea level pressure was 980 mb near the time of 12d. Note the spiraling convective bands.

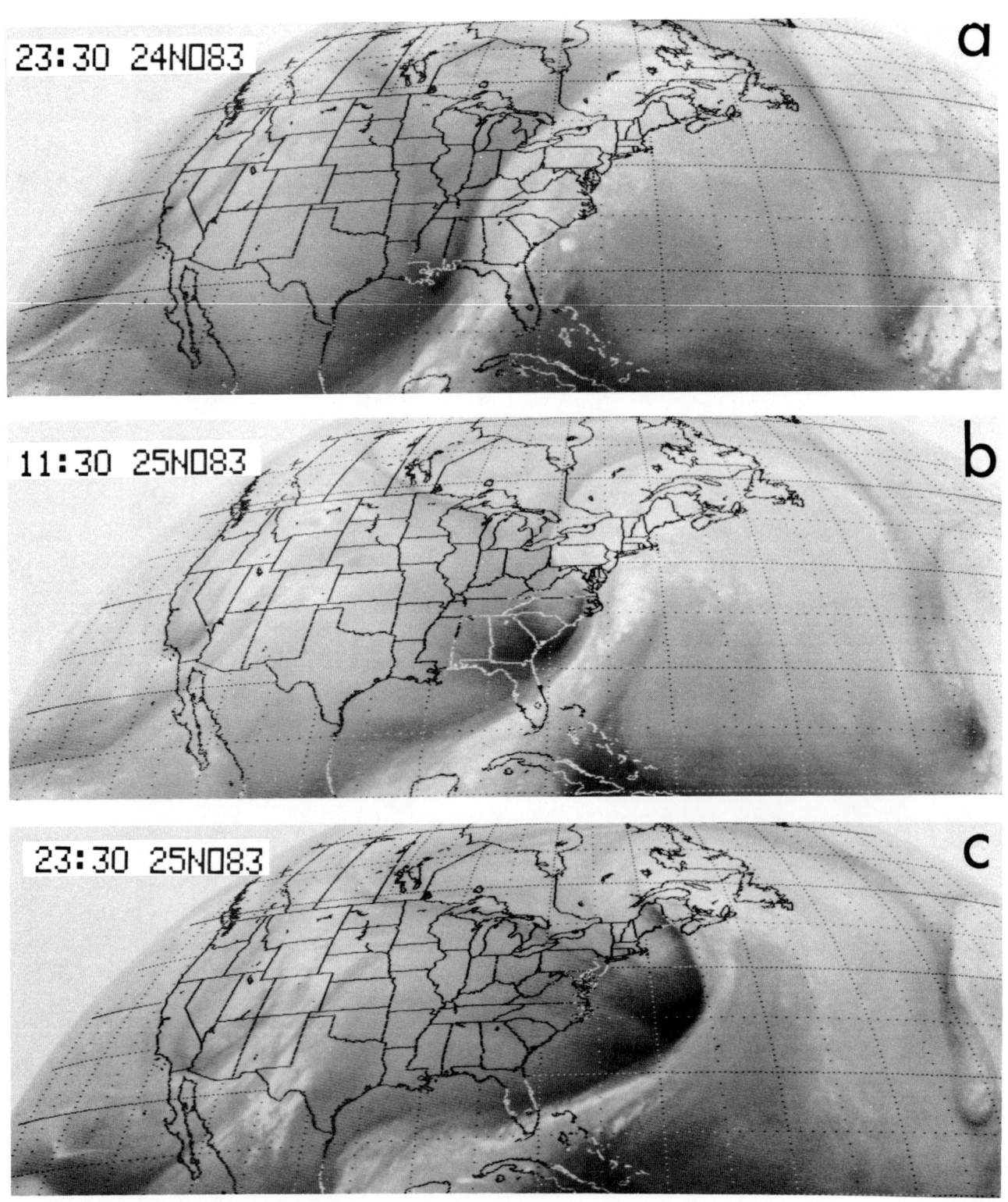

Figures 13a-c. A baroclinic development as viewed in 6.7 μm water vapor imagery on November 24-25, 1983.

Figures 13d-e. Infrared images corresponding to water vapor images in Figures 13a and 13c showing cyclogenesis along the east coast of the United States.

the cloud and moisture imagery off New England. This pattern indicate a low pressure of near 975 mb according to the Junker-Haller technique for estimating surface pressures. The actual surface pressure reported was 972 mb.

As another example, on March 4, 1984, a slight bulge in the water vapor pattern was noted on the 0515Z 6.7 µm image near 39°N 175°E (Figure 14). By 1615Z, a pronounced "S" shaped area was located near 170°W with a pronounced dark band. The corresponding infrared image for 1915Z March 4 shows a positive vorticity advection area between 30°N to 35°N and 165°W. The 6.7 µm imagery shows a well defined dry slot, indicating continued deepening of the system at 0515Z March 5. By 0415Z March 6, the system was located near 45°N 152°W and had deepened to its maximum potential according to satellite derived parameters, with an indicated surface pressure in the 980 mb range. The actual pressure according to the NMC surface charts was 985 mb.

The satellite imagery for nine hours later at 1315Z shows the system apparently beginning to fill near 50°N 155°W as the frontal band has become eroded. The moisture channel imagery for this same system at 1615Z indicates further weakening by the lack of dark bands around the vortex. However, the 6.7 µm data does indicate deepening of the western system near 40°N 160°W as a polar jet feeds into the southern portions of the system. This proved to be the case as a large low was located near 40°N 150°W by 12Z on March 7.

8. Summary

Cloud pattern evolution observed in satellite pictures can be used to help a forecaster decide whether or not cyclogenesis is occurring. It can also provide estimates of the central surface pressure and at least a qualitative assessment of the deepening rate. Trends observed in moisture channel imagery as well as in the more frequent infrared data appear to lend supporting evidence for cyclogenesis. Further studies are needed to determine more accurately the deepening rate of cyclones, which is critical information for making timely short range marine forecasts.

Acknowledgements

The authors would like to thank Paul McClain and Ralph Anderson for their critical review of the manuscript. Special thanks also go to John Shadid and Gene Dunlap for their assistance with the figures.

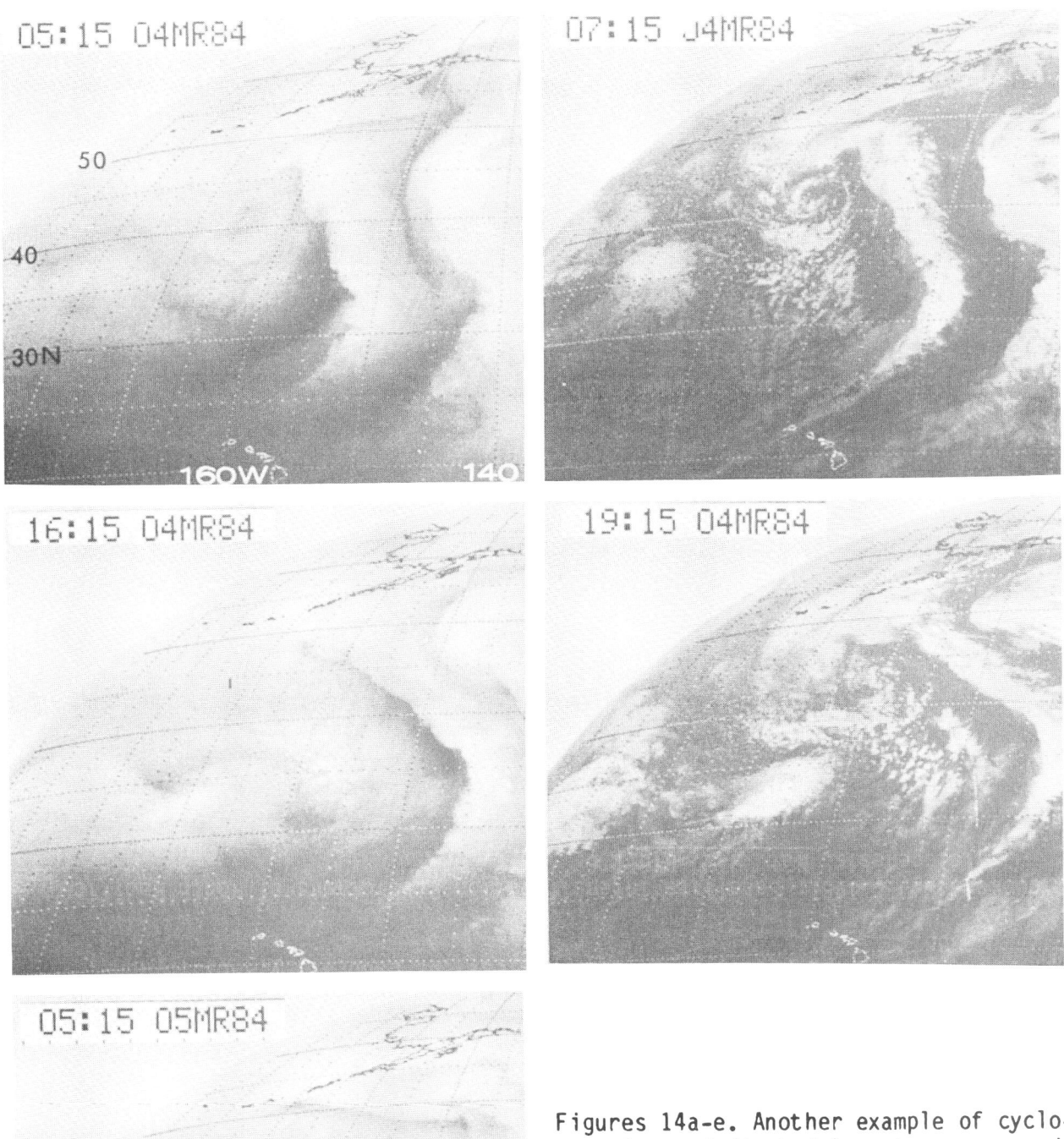

Figures 14a-e. Another example of cyclogenesis as indicated by water vapor and corresponding infrared imagery for March 4-5, 1984.

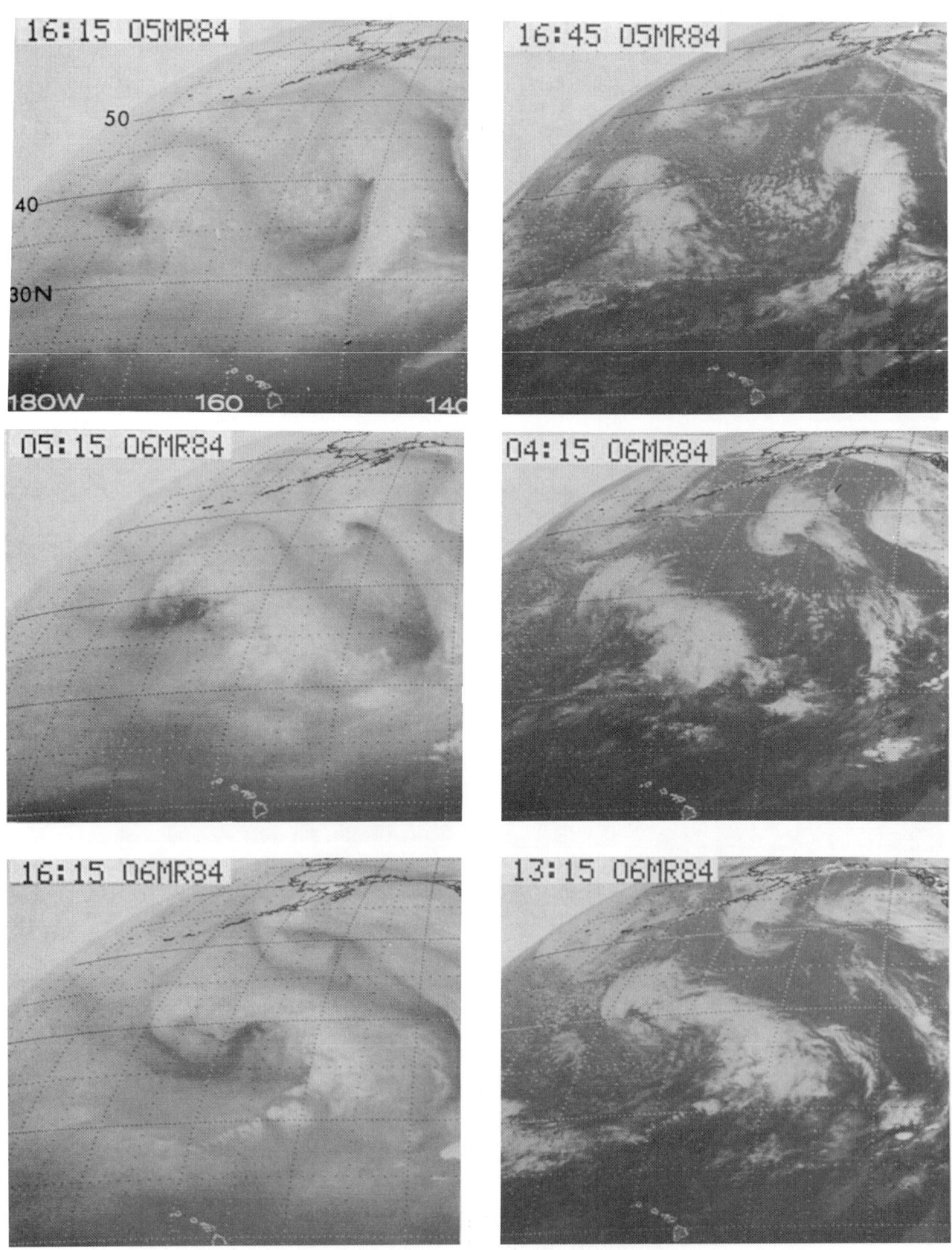

Figures 14f-k. Same storm system as Figures 14a-e for March 5-6, 1984.

References

Anderson, Ralph K., J. P. Ashman, G. R. Farr, E. W. Ferguson, G. N. Isayeva, V. J. Oliver, F. P. Parmenter, T. P. Popova, R. W. Skidmore, A. H. Smith and N. F. Veltischev, 1973: The Use of Satellite Pictures in Weather Analysis and Forecasting, WMO Tech Note #124, Geneva, Switzerland, 154-161.

_____, V. J. Oliver, E. W. Ferguson, F. E. Bittner, A. H. Smith, J. F. W. Purdom, G. R. Farr, and J. P. Ashman, 1974: Applications of Meteorological Satellite Data in Analysis and Forecasting, ESSA Tech Report, NESC #51, U.S. Dept. of Commerce, Washington, D.C.

Burtt, T. G. and N. W. Junker, 1976: A Typical Rapidly Developing Cyclone as Viewed in SMS-2 Imagery, Mon. Wea. Rev., 104, 489-490.

Carlson, T.N., 1980: Airflow through Midlatitude Cyclones and the Comma Cloud Pattern, Mon. Wea. Rev., 108, 1498-1509.

Guymer, L.B., 1978: Operational Application of Satellite Imagery to Synoptic Analysis in the Southern Hemisphere, Technical Report 29, Dept. of Science, Australian Bureau of Meteorology, 83 pp.

Hubert, L. F. and L. F. Whitney, 1971: Wind Estiamtion from Geostationary Satellite Pictures, Mon. Wea. Rev., 99, 665-672.

Jager, G. L., 1984: Satellite Indicators of Rapid Cyclogenesis, Mariners Weather Log, 28, 1-6.

Junker, N. W. and D. J. Haller, 1980: Estimation of Surface Pressure by Satellite Cloud Pattern Recognition, Proc. Eighth Conf. on Weather Forecasting and Analysis, June 10-13, 1980, Denver, CO., Amer. Meteor. Soc. 119-122.

McClain, E.P. and H.J. Broderick, 1967: Recent Research on the Application of Meteorological Satellite Data to Numerical Weather Analysis, AWSTR 196, 45-50.

Rasmussen, E., 1981: An Investigation of a Polar Low with a Spiral Cloud Structure, J. Atmos. Sci., 38, 1785-1792.

Sanders, F. and J.R. Gyakum, 1980: Synoptic-Dynamic Climatology of the "Bomb", Mon. Wea. Rev., 108, 1589-1606.

Smigielski, F.J., 1973: Idealized Development Patterns of Surface Cyclogenesis, Unpublished technical note, Synoptic Analysis Branch, NESS, Washington, D.C., February, 1973.

Troup, A.J. and N.A. Streten, 1972: Satellite-observed Southern Hemisphere Cloud Vortices in Relation to Conventional Observations, J. Appl. Meteor., 11, 909-917.

Weldon, R., 1979: Satellite Training Course Notes, Part IV, Cloud Patterns and the Upper Wind Field, Satellite Applications Laboratory, NESDIS/NOAA, 79 pp.

Younker, W.J., 1981: Satellite Intensification Predictions of North Pacific Cyclogenesis, Nat. Wea. Dig., 6, 40-47.

U.S. DEPARTMENT OF COMMERCE

National Weather Service/National Environmental Satellite Service
SATELLITE APPLICATIONS INFORMATION NOTE 2/77

FIRST GLANCES CAN BE MISLEADING
IN LOCATING VORTICITY CENTERS

Carl E. Weiss
Applications Group, NESS

A problem in both static and animated IR picture interpretation is that an apparent center of cloud rotation frequently does not coincide with a center of maximum vorticity.[1] This is particularly true with rapidly moving short-wave systems in the westerlies. According to Weldon (1976), a cloud rotation center is defined as "an apparent 'center', 'axis', or 'hub' of rotation within the cloud motions when viewed in time lapse motion." The GOES-1 IR imagery on 3 September 1976 illustrates such an example.

At 0001 GMT on 3 September, cloud rotation about (R), Figure 1, was evident in the IR movies. The LFM vorticity analysis at 0000 GMT overlaid on this view, Figure 2, shows the vorticity center (X) to be farther west than the cloud rotation center (R). This is not an obvious location in view of the cloud pattern. The cloud rotation (R), in this case, lies ahead of the vorticity center and is to the north of mid-level wind maximum (S to S') in the area where the vorticity is strongly cyclonic. The cloud rotation center (R) becomes better defined in the IR imagery at 1200 GMT, Figure 3. Here the curvature (C) of the accompanying cirrus to the north of the center (R) has become more pronounced. At this time, a closed 700 mb low (L) was analyzed to the northwest of the rotation center. Examining the 1200 GMT LFM vorticity analysis with the concurrent IR imagery, Figure 4, one can see that the vorticity center (X) continued to be to the west of the cloud rotation (R) in an area relatively free from any middle or high cloudiness. As before, the cloud rotation is found just north of the leading edge of the wind maximum (S to S') where the vorticity gradient is strong.

This apparent discrepancy between the locations of the vorticity center and the cloud rotation center is common. Several possible explanations for this disagreement come to mind. First, since systems often slope in the vertical, clouds showing rotation in the satellite imagery may not coincide with the analyzed vorticity center at 500 mb. However, studies of systems with cloud tops near 500 mb revealed that a similar spatial difference between the location of maximum vorticity and rotation centers still existed. Secondly, a systematic bias in the vorticity analysis could also lead to this same discrepancy. Typically, only over oceans or at other locations where cumulus congestus or cumulonimbus are found behind the comma cloud, do the vorticity centers and the cloud rotation centers agree. In those cases over the oceans, clouds are sometimes present at the vorticity center and therefore the maximum

---

1. In this discussion, a maximum vorticity center is defined as a point at a given level where the value of absolute vorticity is at a maximum. At 500 mb, this correspondends to an "X" on an NMC vorticity analysis.

turning is seen there. More commonly, however, the vorticity center is free from any significant cloudiness. It is obvious, but often overlooked, that cloud rotation can only be seen where clouds are present. Therefore, cloud rotation is frequently found at some location other than the vorticity center. Weldon (1976) depicts a typical short-wave comma cloud pattern with the corresponding vorticity and streamline analyses, Figure 5. The vorticity center (X) is just to the rear of the comma head (A). If rotation in the clouds is evident, it will most likely be in the hatched area (B) along the shear axis of vorticity (dot-dash line, D to D'). This relationship between the vorticity field and the cloud pattern is common for most short-wave troughs imbedded in a strong westerly flow. In summary, by recognizing the typical relationship between cloud rotation centers as seen in satellite imagery with analyzed vorticity centers, errors made in locating vorticity centers (which can be more serious than no analysis at all) can be minimized.

REFERENCE

Weldon, R. B., 1976, unpublished satellite training notes. (Available from NESS, Applications Group (S121), World Weather Building, 5200 Auth Rd., Washington, D.C., 20023.)

Errata; SAIN 76/23: Last sentence, first paragraph should read: A flash flood watch had been issued for the nighttime hours before the event occcurred and had been extended for Western Maryland for the period of flood-producing rains.

Figure 1: GOES-1 Full-Disk IR Imagery, 0001 GMT, 3 September 1976

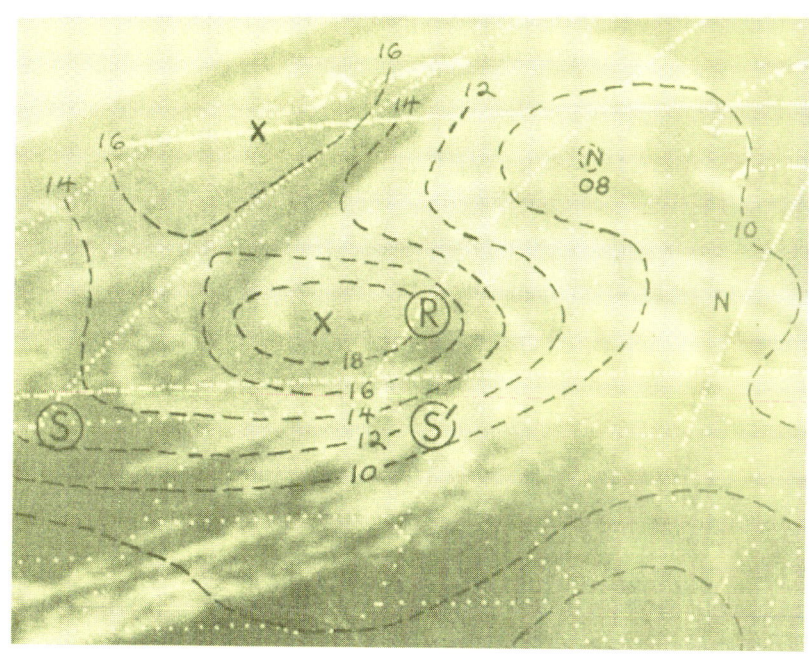

Figure 2: GOES-1 Full-Disk IR Imagery, 0001 GMT and NMC LFM Vorticity Analysis 0000 GMT, 3 September 1976.

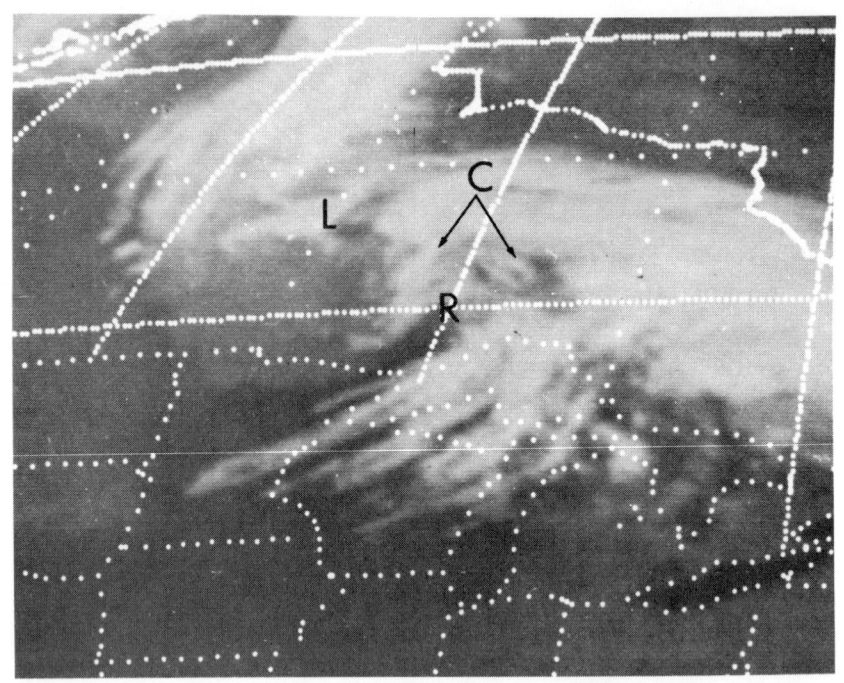

Figure 3: GOES-1 Full-Disk IR Imagery, 1200 GMT, 3 September 1976

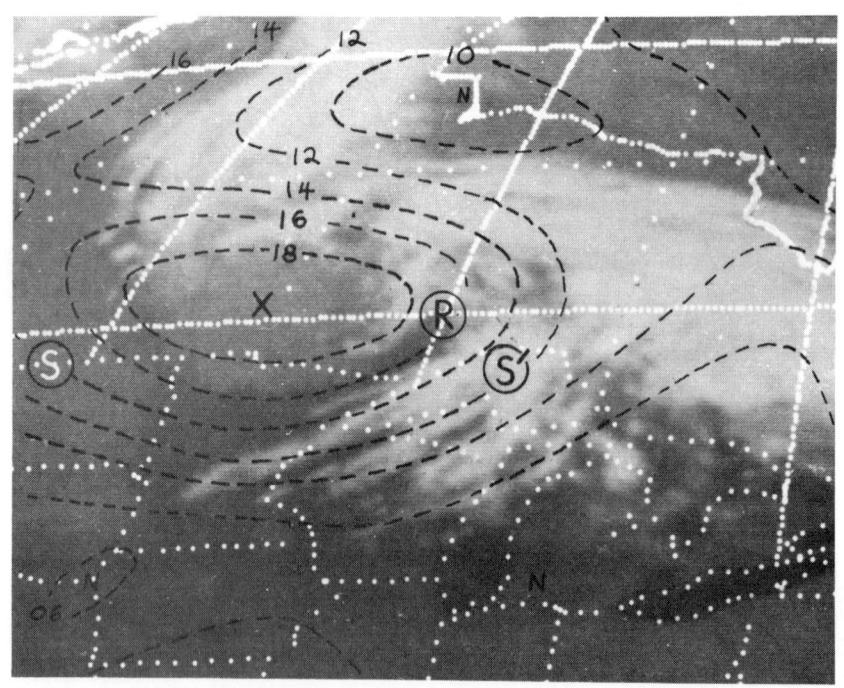

Figure 4: GOES-1 Full-Disk IR Imagery, 1200 GMT and NMC LFM Vorticity Analysis 1200 GMT, 3 September 1976.

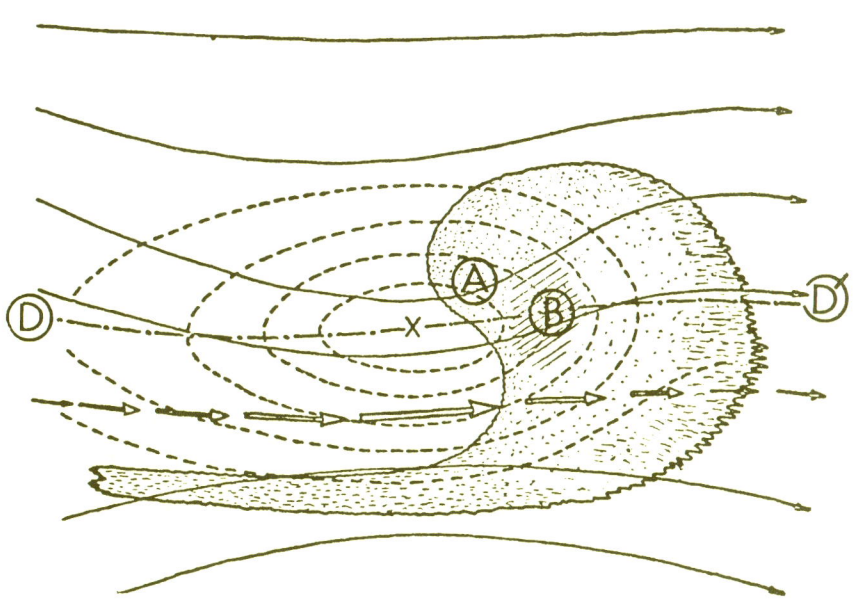

Figure 5: Schematic showing the relationship between the cloud pattern, streamline pattern, vorticity pattern and jet stream in a short-wave system in the westerlies.

U.S. DEPARTMENT OF COMMERCE

National Weather Service/National Environmental Satellite Service
SATELLITE APPLICATIONS INFORMATION NOTE 77/7

AN OCEANIC CYCLOGENESIS - ITS CLOUD PATTERN INTERPRETATION

R. B. Weldon
NESS, Applications Group

Shown is a series of 13 infrared pictures over the Pacific. The pictures are at 3-hour intervals and span a total period of 26 hours. Picture 14 is a visible image which corresponds to the time of picture 6 of the IR series. The cloud pattern evolution shown is a type which is often observed with relatively rapid cyclogenesis over the oceans. Many aspects of the cases are pertinent to weather systems that occur over North America as well.

Note the cloud pattern that is labeled "A" on pictures 1 and 3 and its evolution during the 36 hour period. Some aspects of the cloud pattern and its evolution are unique to the case shown; other aspects are common to many systems of the type. It is those common aspects that are emphasized in the following observations and interpretations.

The cloud pattern evolution:

Most of the cloud pattern change occurred during the 24 hour period which began at 12 GMT, 3 February, and ended at 12GMT, 4 February (pictures 3 thru 11). On picture 3, the cloud pattern - labeled "A" - is well defined with a relatively uniform cold top area and a distinct edge on the poleward side. The "distinctiveness" of the cloud pattern has increased during the previous six hours (compare pictures 1, 2 and 3), although the general "wing" or "leaf" shape has continued. The cloud top area has become more uniformly cold and the poleward edge has sharpened - become better defined. In many cases, this trend is accompanied by a clearing or decrease of cloudiness adjacent to the main cloud pattern, but this is not apparent on the pictures shown here. I have noted that such an increase in "distinctiveness" is an indication that the system is intensifying; the mid-tropospheric vorticity and temperature gradients are increasing with time, and the value of the maximum vorticity associated with the cloud system is probably increasing with time.

Note that the poleward edge of the cloud pattern on picture 3 has a low amplitude "S" shape. On FIGURE 3 - which contains the same IR image as picture 3 - this edge is identified by the white letters "a", "b" and "c". The shape of the edge reverses from slightly concave to slightly convex at point "b", an inflection point. On picture 4, that portion of the edge which was concave west of the inflection point (section "a-b" on Figure 3), now appears as a separate cloud band. The clouds which were south of the edge in that portion of the system have begun to dissipate - or have stopped forming. This trend continues. On picture 6, the edge band itself is no longer detectable; and the cloud pattern has begun to form a comma-like shape.

That part of the edge which was convex east of the inflection point (section "b-c" on Figure 3) has been maintained. This has evolved to become the border of the "head" portion of the emerging cloud comma pattern. On picture 6, a protruding cloud point or "tip" (labeled "D") appears at the rear of the comma "head". When viewed by 30 minute interval time-lapse motion - film loops - the tip appears to have evolved from the inflection point of the original edge. A small clear "slot" appears to the south of the "tip". It is the presence of the "slot" that differentiates between having a "tip" rather than an inflection point at the location where the edge reverses its shape from concave to convex. During the ensuing pictures of the series (pictures 7 thru 13) the slot becomes larger, the portion of the edge which is concavely shaped (south and west of the slot on picture 6) rotates cyclonically with time, and the tip "hooks" cyclonically around the center of maximum vorticity.

Picture 14 is a visible image which corresponds in time to the IR image of picture 6. This was the first high sun angle visible picture available to me for that date. Note the bright distinct pattern and the smooth top appearance of the main cloud system on the visible picture. Such distinctiveness on both the visible and IR images is an indication that the cyclogenesis is occurring through a vertically deep layer. With such a case over an oceanic environment, it is likely that significant surface cyclogenesis is accompanying the upper level cloud system development. Had there been visible data available for the time of picture 3 (about midnight local time), cloud pattern "A" would likely have displayed a similar bright well defined pattern as on picture 14. This opinion is based upon observations of other similarly developing systems at the same phase of their evolution.

The cloud system "movement" and relationships to the wind field:

Which way will it move? If, at the time of one of the first 3 pictures, cloud system "A" had been recognized as an incipient pattern of a developing storm system, this would have been one of the basic questions asked. Another related question is: "which way did it move?" The first requirement for answering either of those questions is to define "it". Figure 4 is a drawing which depicts the movement of the edge "a-b-c" which was indicated on Figure 3. The edge position is shown at six hour intervals. The numbers correspond to picture numbers. Point "a" was defined as the location where the cirrus which was coming over the upstream high level ridge crossed the "tail" end of cloud system "A". Point "b" was defined as the inflection point of the original edge, which became the "tip" as the system evolved. Point "c" was defined as the point on the "outflow" end of the edge where the clouds either ended or became ragged. Keeping in mind that picture 9 was a 0615 GMT picture- 30 minutes later than it should have been to maintain the 3 hourly intervals, the following observations are made:

Note that prior to picture 5 the cloud system had changed little in shape and moved rapidly toward the east-southeast. It was during the period between pictures 4 and 6 that the "a-b" section of the radar edge became banded and disappeared - shifting that part of the edge southeastward and forming the comma-like shape. It was during the same time period that the inflection point/tip part of the cloud system changed from moving eastward or slightly south of eastward to northeastward. Both the change of direction and the transition of the cloud pattern are coincident with the development of southwesterly winds in advance of the vorticity center at mid tropospheric levels. A "waving" or increase of amplitude of the streamlines at middle levels occurred, and the system began to "turn the corner".

Studies of such cloud systems over the U.S. data network have shown that the "vorticity center" - or the point of maximum absolute vorticity value of the 500 mb winds - is located just to the rear of the inflection point on the clear side of the distinct edge - as seen on pictures 1,2, or 3. On picture 6 it would be located southwest of the "tip" feature. As the pattern evolves and the amplitude of the "trough" in the wind field and height contours increases, the vorticity center moves faster than the "tip". I have placed dots on Figure 4 indicating where I believe the vorticity center would most likely be located with respect to the cloud pattern at each of the 6 hour intervals. Note that a line drawn between the vorticity center and the "tip" (or the inflection point of the edge) at each picture time would rotate cyclonically as the pattern evolved. This changing relationship is directly related to the fact that the system was developing an increasing amplitude trough (and eventually a closed low and streamline center) and to the fact that the vorticity center (and the tip feature and comma head) were moving in a path that curved increasingly to the left with time.

In regard to the part of the system that moved northeastward, another point to consider is: can we determine from the satellite pictures when the upper level center "closed" in the contours and streamlines - if it did? Consider picture 7. If we examine the cloud motions on movie loops at the time of that picture, or those of other comma shaped cloud systems in a similar phase of development; the following is observed. The small cloud elements on the north or northwest side of the slot are moving backwards - or losing ground - with respect to the overall cloud system movement. Clouds are dissipating at the edges of the tip. On the other side of the slot, clouds seem to be forming. In other words, the air northwest of the slot is losing ground to the moving cloud system, and the air southeast of the slot - and along the edge southward of that area - is gaining ground with respect to the system. At this phase of development, this differential motion is due mostly to wind shear and only slightly to curvature. The air northwest of the vorticity center - at cloud levels - is still moving from the west or southwest with respect to the map, but not with respect to the system which is moving faster. A "Closed" circulation is present in the relative motion of the air, but not in the winds. As the system evolves and the circulation "closes" in the wind field, the following behavior is observed on movie loops. Cloud elements move back along the tip area and dissipate; those that move furthest before dissipating define the end of the tip. There is considerable variation in this process at half hourly intervals, and a corresponding variation in the shape and extension of the tip. As the system evolves, these elements tend to move more to the left and further along the path than their predecessors before they dissipate. The result to the pattern is, that the tip begins to hook cyclonically around some point moving with the system. In the case shown, this pattern appears on picture 9 and is better defined by picture 10. Experience has shown that this process of the tip hooking around into a curved pattern occurs at about the same time that the elements begin moving back with respect to the ground (as well as with respect to the system). Thus, the hooked tip pattern is a good indication that the system is "closed" in the wind field - and contour field - at cloud levels. Careful study of the film loops for the case shown indicates that the system was "closed" in the cloud level flow: probably by picture 9 and definitely by picture 10. Note that the tip can be hooked completely into a circle; and, if the system is still moving, the vorticity center will not be in the center of the circle formed by the hooked tip. It will be offset to the right of the center of the hooked circle looking downstream in the direction of system movement.

Whereas the comma "head" portion of the cloud system turned northeastward, the part of the cloud edge indicated by "a" continued to move eastsoutheastward until picture 8, then eastward until picture 10. Section "a-b" of the edge rotated cyclonically, such that point "a" which was about 500 miles westsouthwest of "b" on picture 1 is nearly the same distance to the southsoutheast of "b" by picture 13. Edge section "a-b" forms the rear edge of the "tail" cloud feature of the evolving storm cloud "comma". The tail cloud band, in this case, indicates a vertically deep frontal zone extending thru the middle and lower troposphere. Significant convection occurs along that frontal zone from the comma head southward (or southwestward on the earlier pictures) to point "a" where the cirrus crosses over the frontal zone. The northern edge of the cirrus which crosses the tail/frontal zone is near to and parallel to the axis of maximum winds at high tropospheric levels. Thus, point "b" represents the point where the axis of the jet stream crosses the lower level frontal zone. At low levels the frontal zone extends under the jet stream cirrus. The cloud band labeled "f" on picture 14 and on IR picture 6 is likely very close to the location of the surface cold front associated with the developing cloud system.

At the time of picture 9, the southern end of the tail/frontal zone where the jet stream cirrus crosses has been moving eastward, and active convection has been occurring. In the interest of forecasting the weather to the east of the tail at those latitudes, the significant question is: is it safe to extrapolate the past movement of the southern end of the active frontal zone. From experience I have noted that, as long as the cirrus coming over the upstream ridge extends downstream to - or across - the frontal zone tail band as a well defined band or edge nearly parallel to the high level winds, the frontal zone - where the cirrus crosses - will continue to move, and convection will occur along the tail to the left of where the cirrus crosses. When the cirrus edge or band coming over the upstream ridge is well defined, the axis of maximum winds tend to be parallel to the edge and parallel to the isopleth of vorticity. If such an edge is present and nearly straight upstream to the

ridge or back upstream 10 or 15 degrees longitude, then the frontal band
at the crossing point will remain active and move in the direction of the
cirrus that crosses it. I can reliably extrapolate the frontal band
movement for 12 hours or more. If the definition and straightness of the
edge is maintained, but the upstream ridge is building northward with time
the jet and cirrus band will rotate anticyclonically and the crossing point
will move increasingly to the right of its previous track. In this case,
none of the above are occurring. Examination of the clouds of the upstream
ridge indicates two changes occurring during pictures 9, 10, and 11. The
flow between the upstream ridge and the frontal band is increasing its
amplitude. The cloud edge near the ridge is more and more anticyclonically
curved and that near the frontal band is more cyclonically curved. The cirrus
edge near the frontal band is rotating cyclonically in concert with this
increase of amplitude; and the point "a" of the system begins to turn toward
the northeast. Also, as the increase in amplitude occurs, the cirrus no longer
extends from the ridge to the frontal band. The second change occurring is
that the edge near the ridge is becoming ragged indicating that the flow is
across the vorticity isopleths in an "negative vorticity advection" sense. This
also indicates that the strong winds in that area will be spread over a wide
area; the axis of maximum winds will be difficult to define. The first
change indicates that the point of cirrus crossing the frontal band will turn
toward the northeast (or more to the left of its previous track); and the
second change indicates that the system will begin to shear out and weaken.
Note on Figure 5 that the 300 mb contours spread in a different pattern
between the upstream ridge and the evolving storm cloud system. During
the period from picture 11 until picture 13, the cloud system elongated into
a northwest - southeast orientation, and the cloud bands became thinner.
Studies of such cloud system behavior over data rich areas indicate the
following: An elongated region or "lobe" or higher vorticity values extends
from the vorticity center to the point where the cirrus crosses the frontal
band. As the cloud system stretches or elongates in the manner shown here,
the vorticity lobe also elongates, the vorticity gradients decrease, and
the value of the vorticity center decreases.

Because of the change of structure of the upstream ridge, the demise of the
system can be guessed at the time of picture 11. By picture 12, the guess
is reinforced, and by picture 13, the trend is well established. After that
the system continues to "shear out" as it moves against the blocking ridge-
trough pattern over the North American continent.

The early cloud pattern identity - a "baroclinic leaf"

How soon was the development identifiable on the satellite pictures? The
answer is - in my opinion - at the time of picture 1. Although the cloud
pattern was identifiable prior to that time, it was less distinct. By
picture 1, the pattern had a distinct "wing" or "leaf" shape which is characteristic of the incipient cloud patterns in many cases of cyclogenesis. I refer
to such an incipient cloud system as a "baroclinic leaf". I could not be sure
at the time of picture 1 - based merely on the shape alone - that the cloud
pattern "A" was of the "baroclinic leaf" category. However, there are some
additional clues which reinforced the identity.

(1) If the "leaf" shaped pattern is a "baroclinic leaf", it should be located within the jet stream zone, or in the exit region of an advancing jet stream with weaker flow ahead of it. The differences of the low clouds in picture 1 indicate that pattern "A" is within the jet stream zone.

(2) The logitudinal cloud axis of a "baroclinic leaf" pattern will most likely be oriented in a direction which is rotated counter-clockwise by some acute angle with respect to the direction of the upstream jet axis. This is difficult to check on picture 1, since the upstream jet stream is on the horizon of the picture. A related characteristic which can be checked is that the axis of the "leaf" will usually be in a direction oriented counterclockwise from the direction of propagation of the cloud system. This is true at the time of picture 1. The axis of the "leaf" is WSW-ENE, and the direction of propagation a little south of eastward.

(3) On an IR picture, the cloud top area of a "baroclinic leaf" system is usually colder and smoother on the wider portion, with the cloud tops becoming progressively warmer and more irregular toward the narrow "tail-like" end.

From the above combination of characteristics, the "baroclinic leaf" identity of cloud pattern "A" was firmly established; and the interpretation that significant development was underway was a highly reliable one. The primary remaining question is: What is the vertical depth involved in the development? The coldness of the pattern on the IR image indicates that system is affecting the high troposphere, but some "leaf" cloud patterns and their associated development are confined to the higher levels.

In this case, I concluded that the development was vertically deep and probably included the surface. The conclusions were based on the following:

(1) Such coldness on the IR image with respect to other cloud systems on the same picture, implied a vertically thick cloud system.

(2) The jet stream associated with the system appeared to be vertically deep because of the effects on the convective clouds on opposing sides.

(3) The development was not in the southern branch of a split in the westerlies, but appeared to be within the main branch of the jet stream. High level development often occur in the southern branch of splitting westerly jet streams.

(4) The development was over the ocean where a large availability of low level moisture existed.

Another point to check in this regard is:

(5) Does a low level tail-like band of clouds - such as those associated with a surface front - extend to the rear of the narrow end of the "leaf"? If it does the development involves the lower atmosphere. In this case the area where the low band would be is obscured by higher clouds. So, this clue cannot be checked.

Figure 3 shows the 300mb contours from the NMC analysis superimposed on the IR image of picture 3. The purpose of the figure is to illustrate the relationships of this particular "baroclinic leaf" pattern to the long wave environment. Figures 1 and 2 depict "baroclinic leaf" patterns within two other types of environments.

The following conditions are common to "baroclinic leaf" patterns in all 3 types of environments:

(1) The region of the atmosphere under the leaf is frontogenetic. In early stages of most cases, the frontogenesis was strongest in the middle troposphere, and was occurring in the deformation zone in advance of a speed maximum in the jet stream upstream from the leaf. The overall frontal zone is parallel to the longitudianal axis of the "cloud leaf". As the system progresses, lower level frontogenesis begins or increases under that part of the leaf with the concave rear edge. This appears to be enhanced by sinking air to the rear of that part of the cloud system. At high tropospheric levels the frontogenesis is greatest near the convex portion of the edge.

(2) The presence of the "leaf" is an indication that cyclogenesis is likely. If the development is not halted by some change in the environmental conditions, the "leaf" will evolve into the storm cloud comma pattern.

(3) Although the "rear" edge of a "leaf" is often well defined, the winds and height contours at cloud top levels are oriented across the edge at angles ranging from very large at the narrow (usually equator-ward) end to very small or paralled at the wide (usually poleward) end.

(4) The leaf pattern is located within a tight gradient of 500mb vorticity isopleths. The 500mb winds cross the isopleths in a positive advection direction. The vorticity gradient is usually increasing with time. For those systems that are developing rapidly, the vorticity isopleths are parallel to the contours and winds along the jet axis upstream from the narrow end of the "leaf".

(5) The jet stream axis is NOT along the well defined edge of the leaf. The jet stream axis normally crosses the leaf pattern in a manner depicted by the arrows on Figures 1 and 2. The jet stream is usually well defined and strong at middle levels upstream from the leaf where the cross-hatched arrows are shown on Figures 1 and 2. There is more variability in the jet structure on the other side of the leaf.

The following differences which depend upon the 3 types of environments shown:

Figure 1:

Here the "baroclinic leaf" is located over a pre-existing or established low level or surface frontal zone. The clouds associated with the low level front are indicated by the hatched band. As the deep layer frontogenesis occurs, the low level increasing gradient will be merged with the existing gradient of the established surface front along the tail portion of the leaf. In such cases, when the surface cyclogenesis occurs north of the established front, that portion of the newly forming cold front which is north of the established front, will be "an instant occlusion". If the surface cyclogenesis occurs in the surface frontal zone of the pre-existing front, a "frontal wave development" will occur. My observations indicate that the place where the surface cyclogenesis occurs in most dependent upon where the middle and upper level strong wind zones are. The lower level environment - especially the availability of warm moist air - appears to influence the rapidity at which the surface cyclogenesis proceeds. This may be related to the fact that when a "leaf" cloud pattern does occur over or near to an already established surface frontal zone, the convex curved portion of the edge and the clouds in that portion of the "leaf" dominate in coldness and brightness. In fact the cloud system quickly takes on more or an "arch" shape rather than a "leaf" or "wing" shape on the IR images.

Figure 2:

Here the "leaf" occurs on the forward side of a relatively high amplitude trough. Frequently in such cases considerable high cloudiness is present on the warm side of the jet stream axis. Note that the edge of that cloud pattern coincides closely with the axis of the jet stream indicated by the arrows. The edge of the "leaf" does not. In regard to the low level frontal locations, similar variations occur as in the case of Figure 1, since often there is a pre-existing low level frontal zone under the jet stream related cirrus. Frequently a new surface front will form along the tail of the "leaf" before it accelerates eastward and overtakes the colde pre-existing front.

Figure 3:

Here, as in Figure 1, the "leaf" is forming within a relatively low amplitude branch of strong westerlies. However, in this case there is no pre-existing surface frontal zone near the "leaf". Note the southwest to norteast oriented warm cloud bands well to the southeast of the "leaf". These are the remnant frontal cloud bands of previous systems which have moved east and dissipated against the blocking ridge. One of these still had active convection along it in picture 1. In such a case as this - during the early stages of the "leaf" - a surface front may exist, trailing back upstream under the middle and upper level jet/baroclinic zones, or it may form under the tail end of the "leaf" as the system intensifies.

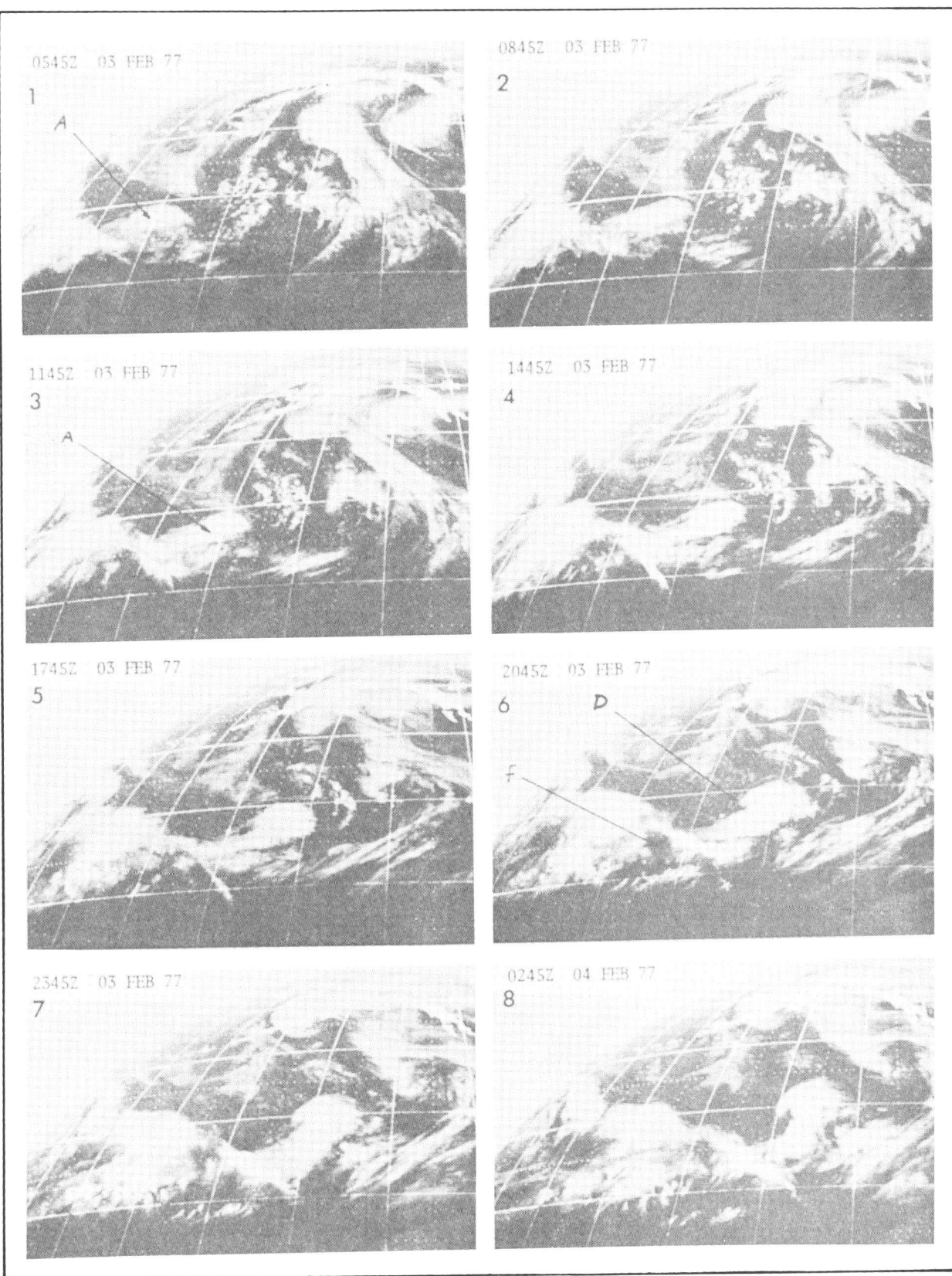

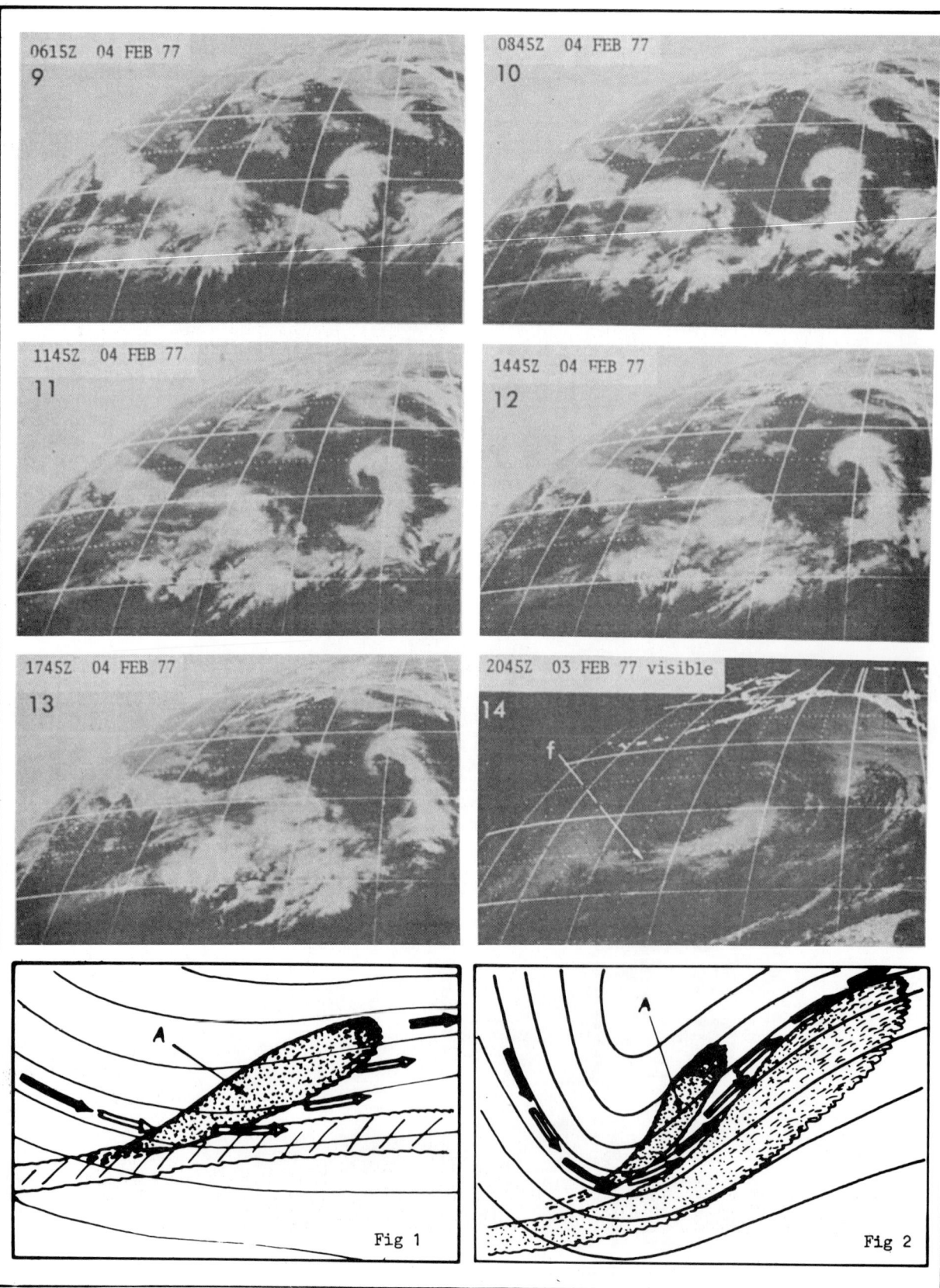

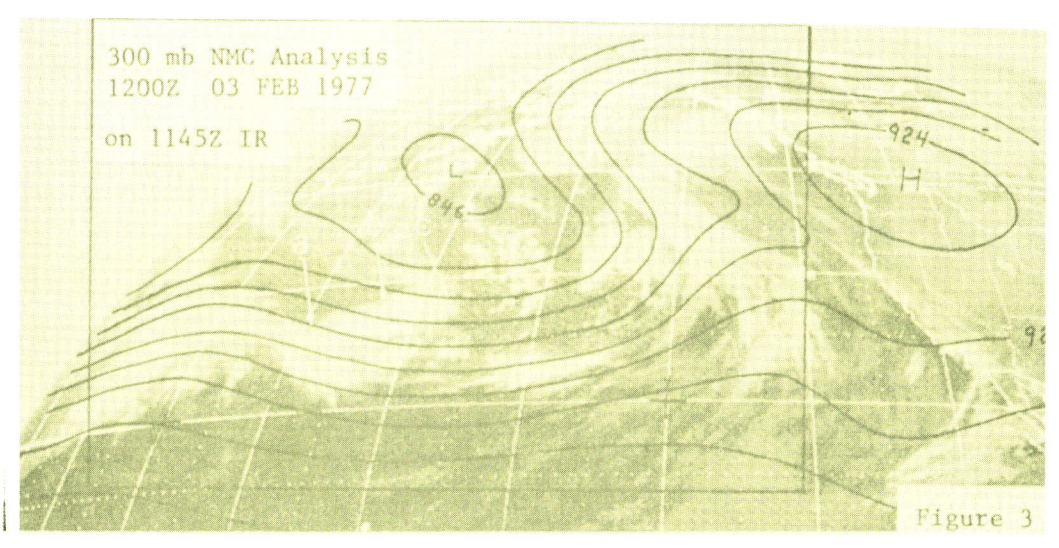

Figure 3

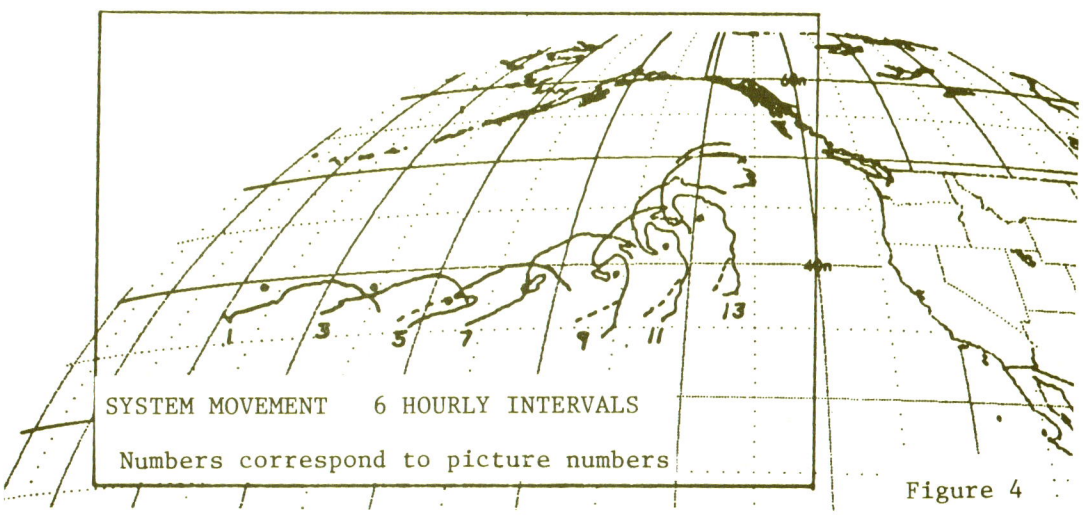

Figure 4

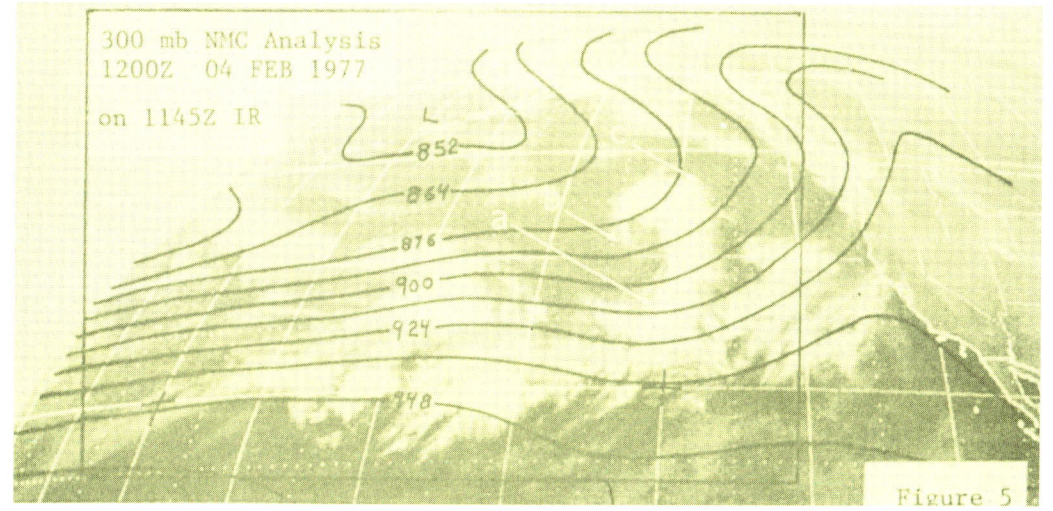

Figure 5

EXTRATROPICAL CYCLOGENESIS OVER THE GULF OF MEXICO

Brian E. Heckman
NOAA, NESS, Satellite Field Services Station
Kansas City, Missouri

A.H. Thompson
Texas A&M University
College Station, Texas

I. INTRODUCTION

One of the more demanding tasks an operational meteorologist has is to predict accurately the time and place of a new extratropical cyclone. The task is even more formidable if the anticipated development is over a data-sparse area such as the Gulf of Mexico. Today, forecasters are relying more and more on output from numerical weather prediction (NWP). It has been shown, however, that over oceanic regions NWP surface pressure forecasts often have significant errors (Leary, 1971).

Early in satellite meteorology, it was shown that extratropical cyclogenesis is often accompanied by observable cloud-pattern changes (Boucher and Newcomb, 1962). In Boucher and Newcomb's five stage cyclone model, the most noticeable cloud-pattern change was widening or bulging of the frontal cloud band.

After infrared (IR) imagery was available, a new dimension in observing cloud changes was available. It was now possible to estimate, qualitatively, atmospheric vertical motion from satellite data. Shenk (1970) observed a noticeable decrease in cloud-top temperatures (CTT) just poleward of where a new cyclone later formed.

After the launch of the SMS/GOES spacecraft, the evolution of cyclogenesis could be observed in much greater detail and at short intervals. Recent investigations using this satellite support cloud-pattern evolution as in the classical Boucher-Newcomb (B-N) model (Johnston, 1974; Cochran and Johnson, 1976 and Fisher, 1977). In addition, the data show the atmosphere is much more complex and variable than earlier models suggest. Weldon (1976) and Burtt and Junker (1976) have observed cloud-pattern evolution associated with rapid cyclogenesis over highly baroclinic oceanic regions.

Investigations of extratropical cyclogenesis in the Gulf of Mexico have been limited. Thompson, et. al. (1965) suggested cyclogenesis followed, with a few exceptions, the classical B-N model.

The objectives of this paper are to focus on the following questions:

*Are there characteristic three-dimensional cloud-pattern changes accompanying Gulf of Mexico development?
*Can classical cyclone development models (such as the B-N model) be used to describe Gulf cyclogenesis?
*If characteristic patterns do occur, can results be utilized in forecasting?

Cases of cyclogenesis during the 1976-77 cool season (October-March) were investigated. Results, along with a case study are given in Section 3.

In Section 4 a preliminary cloud model for Gulf cyclogenesis, based on the 1976-77 cases is given. A discussion of how this model differs from the classical models is given. To broaden the number of cases and test results found in 1976-77, cyclogenesis during the 1977-78 winter season is examined.

2. CRITERIA FOR CASE SELECTION

The primary consideration used in selecting cases was the availability of enhanced IR (EIR) data. Since these data were not available routinely until after the early part of 1976, it was decided to study the entire cool season, 1976-77 (October-March).

To aid in making final selection of cases the following guidelines were used:
(1) Cyclogenesis is defined as the time when a new closed isobar (drawn at 2-mb intervals) and a closed wind circulation can be analyzed when on the previous surface analysis one was not drawn. Primary surface analyses utilized were those from the Miami Regional Center for Tropical Meteorology
(2) The cyclone must have moved onshore so that its existence (after persisting 12 h or longer) could be ascertained.

Cases studied during the 1976-77 season are listed in Table 1.

Table 1. List of cyclone developments, cool season 1976-77, showing certain pressure characteristics (see text for full explanation).

Case	Period Studied	Time of Cyclogenesis	Initial Central Press. (mb)	24h Pressure Change (mb)
1*#	16-17 Oct	16/1200Z	1011	-2
2*#	29-30 Oct	29/1200Z	1012	-5
3*#	14-15 Nov	14/1200Z	1011	-3
4 #	19-20 Nov	19/1800Z	1011	-3
5*#	2-3 Jan	02/1800Z	1011	-5
6	6-7 Jan	06/1800Z	1012	Dissipated
7	24-25 Jan	24/1800Z	1016	6
8	3-4 Feb	03/1200Z	1013	Dissipated

**Average: -4 mb

* Indicates storm persisted 30 h or longer.
**Cases which persisted longer than 24 h.
# Indicates cases with significant satellite signatures.

3 PRESENTATION OF RESULTS

After the eight cases listed in Table 1 were studied, it was determined that five cases were accompanied by a noticeable three-dimensional cloud-pattern change. The remaining cases showed only minor cloud-pattern changes. Since a primary objective of this investigation was to isolate satellite-observed signatures which may be useful in forecasting, emphasis was focused on those cases with the <u>most repetitious</u> signatures. A representative case showing these signatures, the 14 November 1976 case, is presented below.

Figs. 1a-1e show the surface analyses between 0000 GMT 13 November and 0600 GMT 14 November. Fronts and 12-h surface pressure changes are shown. At 0000 GMT on the 13th (Fig. 1a), a well-marked cold front extended across the central Gulf bending sharply near 24°N 93°W then crossing the Mexican coast.

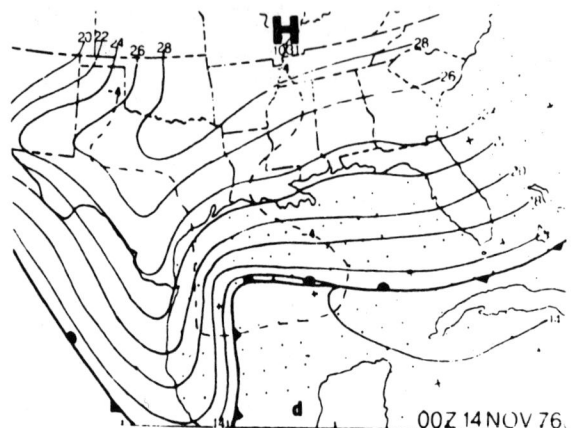

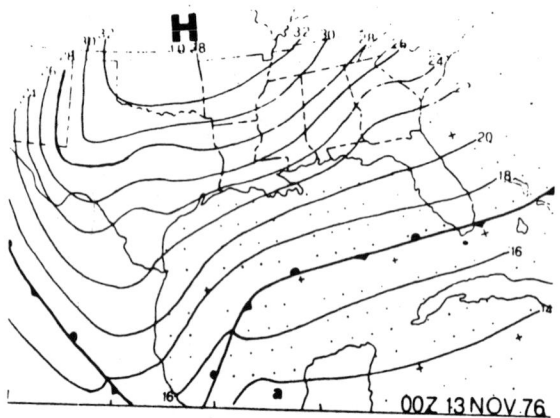

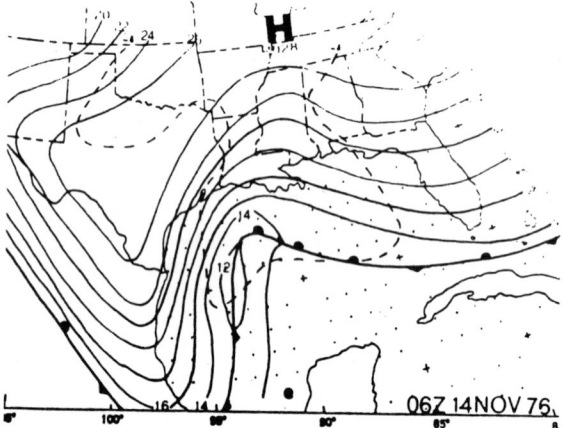

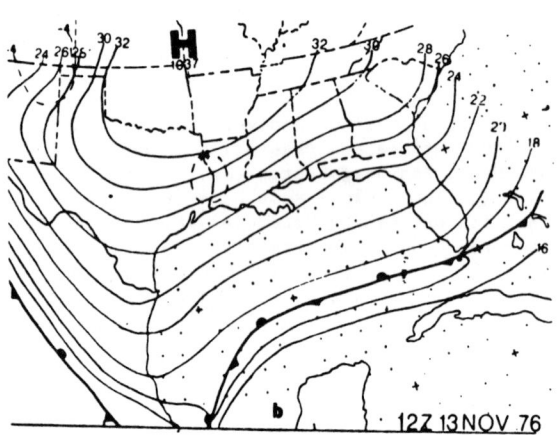

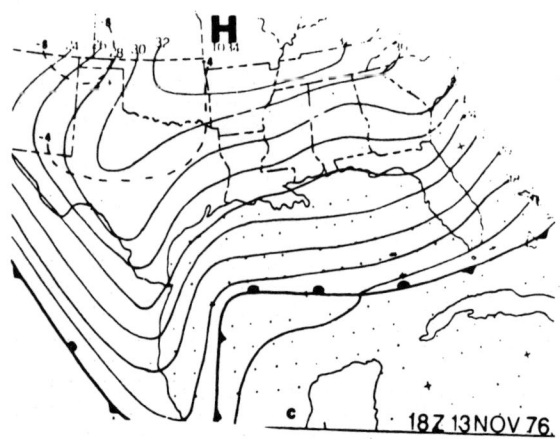

Figs. 1a-1e. Surface analyses with isobars (solid lines) and 12-h isallobars (dashed lines) at 4-mb intervals.

Upper air conditions from 0000 GMT 13 November through 0000 GMT 14 November are shown in Figs. 2a-2c (next pg). The 500-mb analysis at 0000 GMT on the 13th indicated a well-developed southern branch of westerlies which reflects a strong blocking ridge in southwest Canada. Two vorticity maxima*, indicated by x's were moving through this pattern. The associated positive vorticity advection (PVA) areas (labeled A and B) are illustrated by thin solid contouring while absolute vorticity is drawn in dashed lines. The polar jet axis is delineated by the heavy streamline; maximum core speeds in excess of 70 ms^{-1} were observed along the jet axis over Alabama and Georgia. Further south, a segment of the subtropical jet extends from central Mexico northeastward over the southern Gulf coast.

Synoptic conditions at 0000 GMT on the 13th appear favorable for cyclogenesis in the northwest Gulf during the next 24-36 h as the PVA areas moved into juxtaposition with the surface frontal zone. The exact timing for

---

*Vorticity analyses are taken from a computer program developed by the Scientific Services Division, National Weather Service Central Region. Computations are made from observed winds at 500 mb.

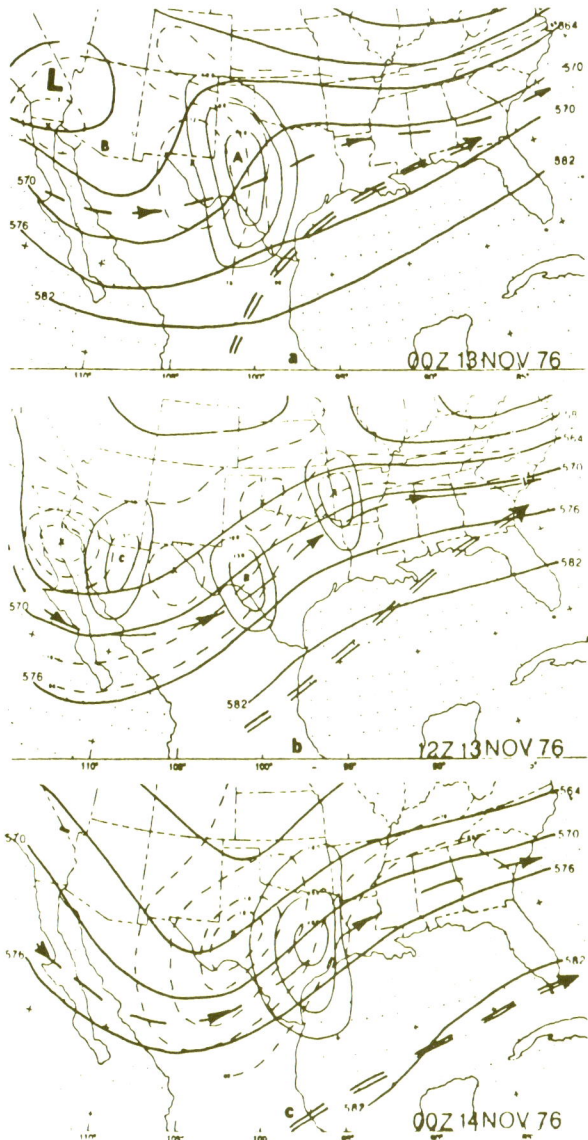

Figs. 2a-2c. Upper air conditions. Heavy lines are 500-mb contours, arrowed line is position of upper-level jet and double-shafted line is subtropical jet. Absolute vorticity at 500-mb (thin dashed lines) in 10^{-5} s^{-1}, positive vorticity advection (thin solid lines) in 10^{-11} s^{-2}. Positions of vorticity maxima are marked by "x".

development, however, was difficult to resolve from conventional data. Numerical guidance for instance, did not forecast development until after upper air data for 0000 GMT 14 November were processed, by which time cyclogenesis had already begun. It is desirable to be able to make a refined, short-range forecast (say 12-18 h) for this event. In this case, satellite information provided an important clue.

EIR data for the same times as the surface analyses are illustrated in Figs. 3a-3e. It may be helpful to explain further the EIR and how inferences of vertical motion are made. Indicated in Fig. 3a are the approximate radiant temperatures (T$_{BB}$ in °C) sensed by the satellite radiometer. If one compares grey shades between two consecutive pictures, one obtains a general trend of how the CTT (or cloud-top height) is changing.

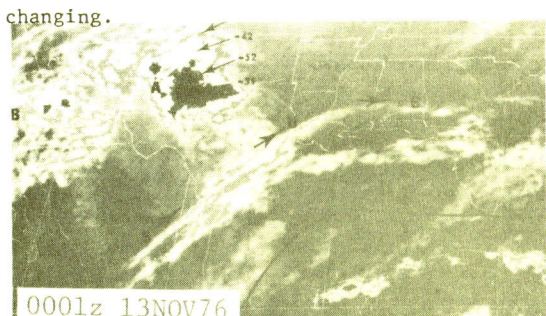

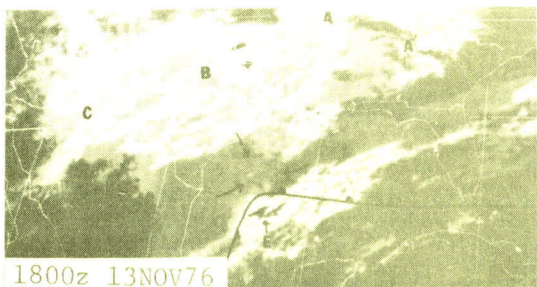

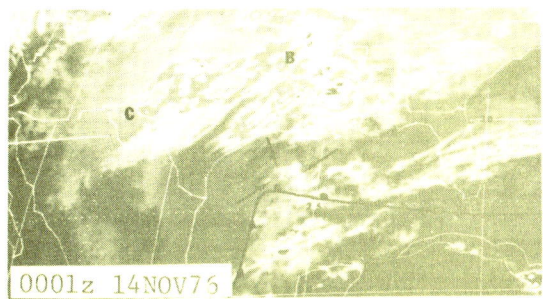

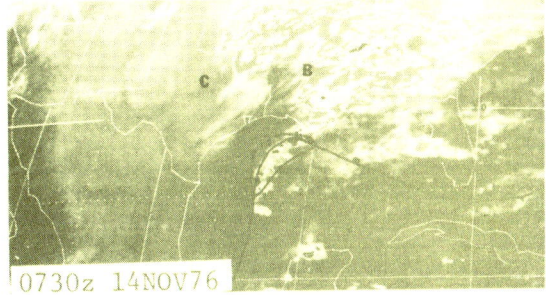

Figs. 3a-3e. Enhanced IR satellite data. See text for explanation of annotations.

Techniques using EIR data to estimate rainfall have found that as the CTT decreases, the cloud is, in general, producing more rainfall (Scofield and Oliver, 1977). This empirical relationship also is applicable in inferring vertical motion i.e., if the CTT decreases, it is assumed there

2-H-3

is upward vertical motion (uvm).

At 0000 GMT on the 13th (Fig. 3a) there is no well marked boundary supporting the frontal position over the Gulf. Best depicted, however, are the clouds associated with the uvm field depicted in Fig. 2a by the PVA areas. Cloud area A correlates well with PVA area A in Fig. 2a. Likewise, the colder cloud-tops in southern Arizona (marked B) are associated with PVA B. Also marked well, is the segment of the subtropical jet (S-U-B).

During the next 12 h (1200 GMT 13 November) PVA area A moved northeastward. Upper divergence ahead of this shortwave trough had a negligible effect on the surface pressure pattern, as no 12-h pressure falls developed (Fig. 1b).

Although development did not occur, conditions remained favorable for cyclogenesis as two additional vorticity maxima were evident in Fig. 3b. One is located in west Texas (compare PVA B in Fig. 2b with B-B' in the EIR data). The second vorticity maximum, labeled C, is the pattern associated with the upper trough in southern Arizona.

After 1200 GMT, cyclogenesis progressed rapidly and was reflected in the EIR data as a noticeable change in the horizontal and vertical extent of clouds near the developing cyclone. As PVA area B (Fig. 2b) moved east-northeastward, upper divergence increased over the northwest Gulf. By 1800 GMT (Fig. 3c) an increase in middle clouds (those not being displayed as the first grey shade) was noted near the arrows. This growth of clouds appears to reflect the uvm field associated with the PVA.

During the next 6 h this area expanded northward with CTT showing a marked decreasing trend (Fig. 3d). Notice also the northward movement of the warm front and the 12-h pressure-fall pattern as the cloud area expanded northward (Figs. 1c-1d).

Continuation of these trends is evident from the EIR data in Fig. 3e. By 0730 GMT convection had developed along the front; this appears to be very near the time of cyclogenesis (Fig. 1e). Convection increased during the next 6 h (not shown) as PVA C moved eastward and increased the upper divergence; the surface low by this time had approached southeast Louisiana (not shown). Within the next 24 h the cyclone moved across the northern Gulf, deepening by another 3 mb.

The events illustrated above show that the horizontal and vertical extent of clouds began to increase noticeably beginning approximately 18 h before cyclogenesis. A good correlation appears to exist between this growth of clouds and the passage of at least two upper-level PVA areas; the uvm fields being the primary mechanism for the observed cloud growth. Somewhat similar cloud changes were observed in the 19 November and 2 January cases (Table 1).

4. DISCUSSION OF RESULTS

It is of interest to look at the overall synoptic features associated with cases which occurred with significant satellite signatures during 1976-77. Salient features occurring approximately 12 h before surface development are shown in Fig. 4a. A single surface front extends through the Gulf; this front separates maritime tropical air (dewpoints $\geq 18°C$) from modified continental polar air to the north.

The mean position of the 500-mb absolute positive vorticity center is in extreme west Texas (indicated with the large dot). An axis of PVA (shaded area) extends southeastward into the northwest Gulf. The polar jet curves cyclonically through Texas then anticyclonically as shown by the streamline.

During the next 12 h (Fig. 4b) the vorticity center moves eastward into west Texas with the PVA axis extending well into the northwest Gulf. The portion of the front near $25°N$ $95°W$ begins to move northward as a warm front with cyclogenesis occurring in the vicinity of $27°N$ $94°W$.

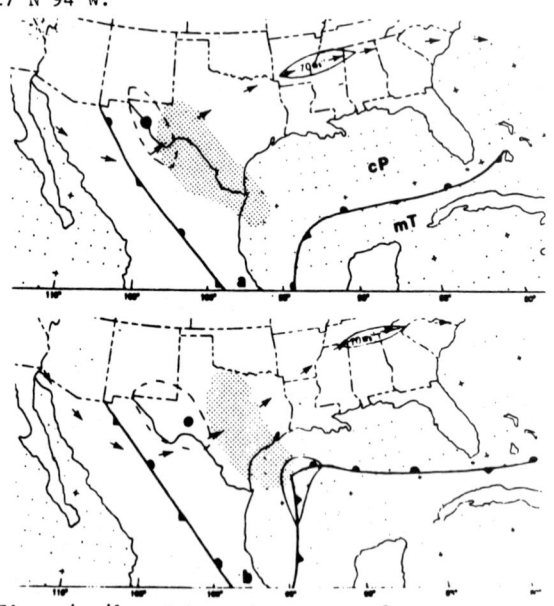

Figs. 4a-4b. Schematic summary for cases (1976-77) of cyclogenesis accompanied by significant satellite signatures for 12 h prior to cyclogenesis (a) and near time of development (b). See text for explanation of annotations.

The most repeated satellite signature during the 1976-77 winter season was a noted decrease in CTT just poleward of the nascent cyclone. Similar signatures have been noted in earlier investigations using IR data from polar orbiting satellites (Sherr and Rogers, 1965 and Shenk, 1970).

Based on three cases during the 1976-77 winter season, a preliminary cloud model of Gulf cyclogenesis is presented (Fig. 5)(next pg). The cloud pattern (these cases would correspond with synoptic conditions as in Fig. 4a) represents conditions as they would look approximately 12-18 h before a closed wind circulation developed at the surface. The area enclosed within the scalloped edges which extends across most of Texas and Louisiana represents clouds related to the uvm field (PVA area) and a strong polar jet. The position of the absolute vorticity maximum is indicated by the large dot.

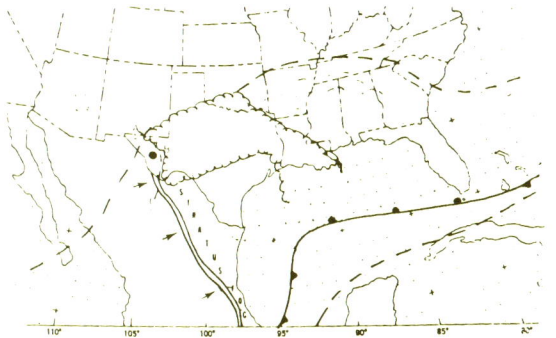

Fig. 5. Preliminary cloud model showing conditions approximately 12-18 h prior to cyclogenesis. See text for further explanation.

The extent of general multi-layered cloudiness can be fairly great and is illustrated by the dashed outline. The double line (also indicated by arrows) in eastern Mexico marks the area where strong easterly low-level flow has packed moisture against the Sierra Madre Orientals and formed a layer of shallow but dense stratus or fog, often very evident in visible satellite imagery. The hatched region just north of the advancing warm front represents the area within which the CTT were observed to decrease most noticeably.

How does this preliminary model compare with previously derived satellite cloud models? Several investigations (Boucher and Newcomb, 1962 and Anderson et. al., 1974) have shown that the most predominate cloud signature accompanying cyclogenesis has been a distinct bulging or widening of a frontal cloud band. This widening is the result of uvm induced by a short-wave trough moving toward the frontal zone.

Unfortunately, fronts in the Gulf seldom exhibit any distinct characteristic prior to cyclogenesis. Deviations from this classical model appear the result of frontal modification specifically characteristic of the terrain surrounding the Gulf and a frequently observed high-cloud pattern emanating from the eastern Pacific.

It is well known that fronts tend to weaken as they push equatorward; this frontolytic process is enhanced if their trajectory is over warm water. After fronts are over the Gulf long enough to become stationary or slow moving, experience shows that they are quite difficult to locate in satellite imagery. Sometimes they are best marked in the visible data as a narrow rope-cloud structure (Jones, et. a., 1976); however, seldom do they exhibit characteristics of a distinct multi-layered frontal cloud band.

Another characteristic of Gulf fronts is the modification of the associated low-cloud pattern by the terrain in eastern Mexico. After moderate to strong fronts enter the northern Gulf, this terrain causes the cold air in the western Gulf to move more rapidly than it does elsewhere along the front. The final configuration of the front is the characteristic bend near 24°N 93°W; a consequence of this behavior is the higher frequency of frontal passages in the western Gulf (Di Mego et. al., 1976).

As the cold air flows west toward the higher terrain, a near-continuous sheet of low clouds is produced by upslope flow which covers sections of southwest Texas, all of eastern Mexico and the western Gulf. The net result of this action is the nearly complete destruction of a straight frontal cloud band that so often is exhibited in other oceanic regions.

A frequently observed cloud-pattern over the Gulf during the cool season is a long fetch of high clouds being advected from the eastern Pacific. This pattern can often be associated with the broad 200 mb trough axis. The subtropical jet axis is sometimes found near the poleward extent of these clouds. If the cloud cover is extensive enough, any lower cloud pattern associated with the front is all but obscured. It was observed that half of the cases studied from the 1976-77 season had, to some degree, this obscuring high-cloud pattern.

5. PRELIMINARY COMMENTS ON THE 1977-78 SEASON

Although a complete investigation of cases during the 1977-78 season has not been completed, preliminary results indicate that cloud-pattern changes were different than in the previous year. The following discussion will focus on the cyclogenesis that occurred on 18-19 January 1978.

The overall evolution of this storm appeared to resemble cloud-pattern evolution noted by Weldon (1976) in his "split-flow" cyclogenesis. Weldon shows four stages of development, the second of which we reproduce as Fig. 6.

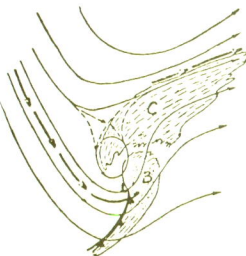

Fig. 6. One of four stages of cyclogenesis under a "split flow" upper-air pattern as viewed in IR satellite data. (After Weldon, 1976.)

The streamlines represent flow at 500 or 300-mb with the jet stream axes indicated by the heavy arrows. The most important cloud development in this type storm as noted by Weldon is the formation of a vorticity comma pattern (cloud area B in Fig. 6). The type of clouds associated with the comma cloud is highly dependent upon the time of year and low-level stability; frequently, however, strong convection has been observed.

Weldon has observed that the relationship between the cloud patterns seen in IR data and the surface low and fronts is highly variable. We have added to the original figure what Weldon stated in his notes to be only a generalization of this relationship.

The cloud patterns shown in the 0430 GMT 19 January EIR image (Fig. 7) resemble those seen in Fig. 6.

Fig. 7. Enhanced IR data for 0430 GMT 19 January 1978. See text for explanation of annotations.

In Fig. 7 the jet stream axis is indicated by the heavy arrows; the positive vorticity maximum is shown by the large dot and surface fronts are in the usual notation. By this time, cloud motions from a satellite movie loop showed a definite twist in the middle and high clouds, suggesting the presence of an upper vorticity maximum cloud (B). In this particular case, the surface low (approximate time of cyclogenesis was 0600 GMT) was displaced about 5° latitude east-southeast of the developing comma cloud.

The convective area northeast of the new cyclone (C) began to develop 16 h before this image. This area appeared to develop as a short-wave trough rotated through the upper trough.

The main difference between cases observed in 1978 and the previous year was the formation of a comma cloud or large vortical cloud over the Gulf. Some insight into possible reasons for these differences may be seen by comparing the overall synoptic situations for the two seasons. In Fig. 8, synoptic features for the January 1978 case are shown.

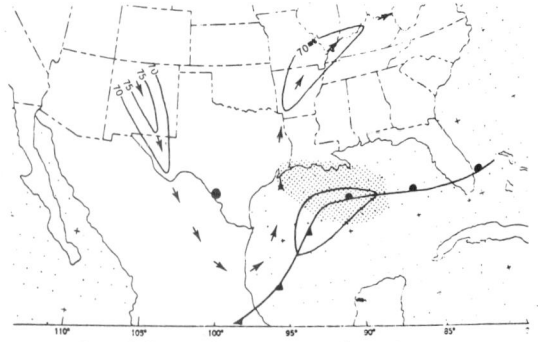

Fig. 8. Schematic summary for the January 1978 case. Annotations are as in Fig. 4.

As can be seen by comparing this to Fig. 4b, several differences exist. First, the jet stream in the 1978 case was displaced much further south and the PVA area is further south. Lastly, surface pressures were about 8 mb lower in the 1978 case, e.g. 1012-mb vs. 1004 in the November and January cases, respectively.

6. CONCLUSIONS

Based on investigations of cyclogenesis during the 1976-77 winter season and a limited review of cases during the 1977-78 season, the following conclusions are offered:

1. Cyclogenesis in the Gulf is frequently accompanied by observable three-dimensional cloud-pattern changes. Cyclones that persist for at least 30 h are likely to evolve with observable satellite signatures. The most prevalent signatures during the 1976-77 season was decreasing CTT just poleward of the nascent cyclone. During the past winter season, the predominant cloud change noted was the formation of a well developed upper vortical cloud pattern.

2. Earlier derived cyclone models (e.g., the Boucher-Newcomb model or slight modifications of it) are inadequate in describing Gulf cyclogenesis.

3. Cloud-top temperatures began to decrease noticeably (in the 1976-77 season) beginning 12-18 h before surface development. This appears to have forecast utility, particularly since these weaker storms are frequently poorly forecast by numerical guidance.

7. ACKNOWLEDGMENTS

The authors would like to express their appreciation to Ms. Sherry Elliott for for the excellent support in typing the manuscript. Mike McDaniel, General Electric, MATSCO is thanked for his expertise in preparing the illustrations.

8. REFERENCES

Anderson, R. K., V. J. Oliver, E. W. Ferguson and F. C. Parmenter, 1974: Application of meteorological satellite data in analysis and forecasting. ESSA Tech. Rep. NESC 51, National Oceanic and Atmospheric Adm., Washington, D.C., viii + 342 pp.

Boucher, R. J., and R. J. Newcomb, 1962: Synoptic interpretation of some TIROS vortex patterns: a preliminary cyclone model. J. Appl. Meteor., 1, 127-136.

Burtt, T. G., and N. W. Junker, 1976: A typical rapidly developing extratropical cyclone as viewed in SMS 2 imagery. Mon. Wea. Rev., 104, 489-490.

Cochran, D. R., and H. Johnson, 1976: Rapid frontal zone cyclogenesis, 31 October 1975. Mon. Wea. Rev., 194, 1078-1080.

DiMego, G. T., L. F. Bosart and G. W. Anderson, 1976: An examination of the frequency and mean conditions surrounding frontal incursions into the Gulf of Mexico and Caribbean Sea. Mon. Wea. Rev., 104, 709-817.

Fisher, A. H., 1977: A satellite view of positive vorticity advection induced cyclogenesis. Nat. Wea. Dig., 2, 27-30.

Johnston, E. C., 1974: Rapid frontal wave development. Mon. Wea. Rev., 102, 804-806.

Janes, S. A., H. W. Brandli and J. W. Orndorff, 1976: The "blue line" depicted on satellite imagery, Mon. Wea. Rev., 104, 1178-1181.

Leary, C., 1971: Systematic errors in operational National Meteorological Center primitive equation surface prognoses. Mon. Wea. Rev., 99, 409-413.

Scofield, R. A., and V. J. Oliver, 1977: A scheme for estimating convective rainfall from satellite imagery. NOAA Tech. Memo. NESS 86, Washington, D.C., 47 pp.

Shenk, W. E., 1970: Meteorological satellite views of cloud growth associated with the development of secondary cyclones. Mon. Wea. Rev., 98, 861-868.

Sherr, P. E., and C. W. C. Rogers, 1965: The identification and interpretation of cloud vortices using TIROS infrared observations. Final report, Contract Cwb 10812, Aracon Geophysics Co., Concord, MA., 77 pp.

Thompson, A. H., M. E. Gosdin and R. F. Jimenez, 1965: Some remarks on cyclogenesis indications as determined from meteorological satellite observations, with application to the Gulf of Mexico.

Weldon, R., 1976: Satellite notes on winter storms. Available from Applications Division, NESS, Washington, D.C.